Rammed earth:
design and construction guidelines

Rammed earth: design and construction guidelines

Peter Walker, *University of Bath*
Rowland Keable, *In Situ Rammed Earth Co Ltd*
Joe Martin, *JM Architects*
Vasilios Maniatidis, *University of Bath*

dti

Details of all publications from
BRE Bookshop are available from:
Website: www.brebookshop.com
or
IHS Rapidoc (BRE Bookshop)
Willoughby Road
Bracknell RG12 8DW
Tel: 01344 404407
Fax: 01344 714440
email: brebookshop@ihsrapidoc.com

Published by BRE Bookshop

Index compiled by Linda Sutherland

Requests to copy any part of this
publication should be made to:
BRE Bookshop
Building Research Establishment
Bucknalls Lane
Watford WD25 9XX
Tel: 01923 664761
Fax: 01923 662477
email: brebookshop@emap.com

EP 62

© Copyright P Walker, R Keable, J Martin,
V Maniatidis 2005
First published 2005
ISBN 1 86081 734 3

Contents

Preface		ix
Acknowledgements		x
1	**Introduction**	1
	1.1 Scope of guidelines	1
	1.2 What is rammed earth?	2
	1.3 Brief history and development	3
	1.4 Advantages and limitations of rammed earth	10
	1.5 Structure of the guidelines	16
2	**Preliminary design considerations**	17
	2.1 Applications	17
	2.2 Influence of rammed earth on other construction activities	22
	2.3 Building control	24
	2.4 Contractual considerations	27
3	**Materials for rammed earth construction**	29
	3.1 Raw materials	29
	3.2 Soil characteristics	31
	3.3 Soil compaction	33
	3.4 Additives	34
	3.5 Soil selection	35
	3.6 Physical characteristics	38
4	**Construction of rammed earth walls**	45
	4.1 Preparation	45
	4.2 Building	51

(continued)

5	**Details for rammed earth construction**	61
	5.1 General	61
	5.2 Footings and base details	61
	5.3 Openings and supports	65
	5.4 Protection given by roofs	69
	5.5 Protective coatings	70
	5.6 Services	74
	5.7 Fixings	75
	5.8 Thermal insulation	75
	5.9 Acoustic separation	75
	5.10 Construction tolerances	78
6	**Engineering design of rammed earth walls**	79
	6.1 Design requirements	79
	6.2 Properties of rammed earth for design	79
	6.3 Simplified design for structural adequacy	81
	6.4 Deformation	84
7	**Maintenance and repair of rammed earth**	85
	7.1 Weathering and deterioration	85
	7.2 Maintenance of rammed earth walls	88
	7.3 Defects in new construction	89
	7.4 Repairs to rammed earth	93
8	**Future of rammed earth**	95

Appendices

	A	Physical properties of rammed earth	99
	B	Specification for rammed earth works	111
	C	Structural wall design	119
	D	Stabilised rammed earth	125

Contact addresses — 131

Glossary — 133

References — 137

Bibliography — 139

Index — 143

Figures

1   Rammed earth wall construction at the Eden Project, Cornwall
2   Construction of a rammed earth wall
3   Rammed earth wall finish, Chapel of Reconciliation, Berlin
4   Traditional rammed earth building, Morocco
5   Seven-storey rammed earth building, Weilburg, Germany (c1820)
6   Rammed earth building, Rhone Valley, France
7   Rammed earth walling at the Alhambra, Granada, Spain
8   Victorian five-storey rammed chalk houses, Winchester, Hampshire (c1840)
9   Victorian rammed chalk building, Andover, Hampshire
10   Rammed chalk house, Amesbury, Wiltshire (c1920)
11   Eden Project Visitors Centre, Cornwall
12   AtEIC Building, Centre for Alternative Technology, Machynlleth, Powys
13   Wall at Chelsea Flower Show 2000
14   Woodley Park Sports Centre, Skelmersdale, Lancashire
15   Rammed chalk walls, Kindersley Centre, Sheepdrove Estate, Berkshire
16   Bird-in-Bush Nursery, London
17   Mount Pleasant Ecological Business Park, Porthtowan, Cornwall
18   Altar, Chapel of Reconciliation, Berlin
19   Rammed earth wall, Brandenburg, Germany
20   Rammed earth wall, Zeesen, Germany
21   Stablised rammed earth house, Rural Studio, Alabama, USA
22   Stablised rammed earth house, Western Australia
23   Dragons Retreat, Devon (stabilised rammed earth)
24   Jasmine Cottage, Norfolk (stabilised rammed earth)
25   Compaction layers in rammed earth
26   Tooled finish in rammed earth
27   Prefabricated rammed earth walls
28   Rammed earth floor
29   Rammed earth floor, Mount Pleasant Ecological Park, Porthtowan, Cornwall
30   Office desk, Engineers HRW office, London
31   Rammed earth wall construction under cover, Centre for Alternative Technology
32   Compaction layers in rammed earth
33   Pneumatic rammer
34   Manual rammer
35   Relationship between compaction moisture and dry density
36   Grading limits for rammed earth soils
37   Propping of walls during drying
38   Traditional timber formwork
39   Cantilevered formwork
40   Australian proprietary static formwork
41   Proprietary concrete static formwork
42   Timber formwork
43   Timber formwork for curved wall

44	Through-bolted formwork
45	Small forced-action screed mixer
46	Pan-style concrete mixer
47	Skid steer loader
48	Rotavator mixer
49	Pneumatic compaction of a stabilised rammed earth wall
50	Compaction using sheeps-foot roller
51	Movement joints
52	Protection of new works
53	Damp-proof course
54	Base details
55	Water damage at the base of a wall
56	Full-height opening between panels
57	Arched opening
58	Opening details
59	Wall plate details
60	Eaves details
61	Peeling failure of sodium silicate protective coating
62	Preferential weathering of sodium silicate treated wall, exacerbated by under compaction
63	Clay plaster, Woodley Park Sports Centre
64	Movement joints in lime render
65	Plan view of embedded electrical services
66	Back box
67	Insulation details
68	Typical vertical movement joint details
69	Limiting thickness for free-standing and supporting walls
70	Simple rules for openings in rammed earth walls
71	Surface weathering from rainfall
72	Concentrated rainwater flow damage
73	Abrasion damage to vulnerable corners in a stabilised rammed earth wall
74	Walls should be protected from other construction activities
75	Colour variation
76	Textural variation in a rammed earth panel
77	Boniness
78	Formwork patterning
79	Surface cracking
80	Patch repair
81	Plucking damage
82	Surface dusting
83	Efflorescence in a stabilised rammed earth wall
84	Genesis Project, Somerset College of Arts and Technology
85	WISE Project, Centre for Alternative Technology, Wales
A1	Shear testing of rammed earth wall panel
A2	Spray erosion test
A3	Abrasion test
C1	Dispersion of concentrated loads
D1	Brimington Bowls Club Pavilion, Chesterfield, stabilised rammed earth
D2	Stabilised rammed earth stables, Ashley, Northamptonshire

Preface

This publication is believed to be a landmark in that it represents the first guidance document for rammed earth construction published in the UK. It has been compiled as part of Partners-in-Innovation project *Developing rammed earth wall construction for UK housing* funded by the Department of Trade and Industry (DTI). The 30-month project has been led by the University of Bath and In Situ Rammed Earth Co Ltd, working together with Engineers HRW, JM Architects, Knauf Insulation and Mark Lovell Design Engineers as contributing industrial partners. Advisory steering group members included representatives from Bristol City Council, BRE, Day Aggregates, The Ecology Building Society, Feilden Clegg & Bradley Architects, International Heritage Conservation and Management, Grimshaw Architects, Simmonds Mills Architect-Builders and Somerset Trust for Sustainable Development.

The project has included an experimental investigation of material properties, including thermal conductivity testing, structural testing of walls and columns, a worldwide review of rammed earth construction publications and a pilot case study project. As a result we believe that these guidelines represent the current state-of-the-art best practice in rammed earth construction as applicable to the UK. We hope that they will promote and lead to a greater use of rammed earth wall construction and encourage its future development. We welcome feedback and comments for future editions. Finally, we wish to express our sincere thanks to all who have helped to make this publication a reality.

Peter Walker
Rowland Keable
Joe Martin
Vasilios Maniatidis

Acknowledgements

The authors thank the DTI Partners in Innovation scheme for supporting this project. Contributions from the following partners are gratefully acknowledged: BRE, JM Architects, Mark Lovell Design Engineers, Engineers HRW, The Ecology Building Society, Feilden Clegg & Bradley Architects, Knauf Insulation, Grimshaw Architects, International Heritage Conservation and Management, Simmonds Mills Architect-Builders, Bristol City Council, Somerset Trust for Sustainable Development, and Day Aggregates. Special thanks to the following individuals whose comments on various drafts have been extremely helpful in the compilation of the guide: Jenny Andersson, Dirk Bouwens, Dave Clark, Jörg Depta, Stephen Dobson, Matthew Hall, Toby Hodsdon, Chris Massie, Tom Morton, Gordon Pearson, Martin Rauch, Bill Swaney, Steve Vary and Colin Williams. Finally, thanks to Jon Shanks for preparing drawings.

All photographs were taken by Peter Walker unless otherwise stated.

1 Introduction

1.1 Scope of guidelines

For most building designers, rammed earth walling is a novel, innovative and unfamiliar material and construction technique. These guidelines have been compiled with the specific aim of informing, developing and promoting the use of rammed earth wall construction in the UK as a high-quality and sustainable building technology for walls in housing and other low- and medium-rise buildings. Specifically, the guide seeks to encourage the greater use of rammed earth, free from additives such as cement, as an alternative, sustainable and beautiful wall building material.

These guidelines for rammed earth cover general design considerations, material properties, testing and selection, engineering design, wall construction, construction details, and maintenance and repair procedures. A glossary, reference list and bibliography are also included.

> **Note on stabilised rammed earth**
>
> Stabilised rammed earth is an alternative form of wall construction that uses the rammed earth technique, but includes cement, primarily as an additive to change the material's physical characteristics. Stabilisation enhances material durability and wet strength, but at the expense of using cement, a major contributor to global CO_2 emissions. Much of the guidance given here for rammed earth construction is applicable to stabilised rammed earth as well. Where the approaches differ, in material selection for example, these variances are briefly outlined in Appendix D. Further guidance on stabilised rammed earth is also available elsewhere[1,2,3].

1.2 What is rammed earth?

Rammed earth is a form of unbaked earthen construction used primarily to build walls. Other applications include floors, roofs and foundations. Recently it has also been used for furniture, garden ornaments and other features. Rammed earth is formed by compacting moist sub-soil inside temporary formwork (Figures 1 and 2). Loose moist soil is placed in layers 100–150 mm deep and compacted. Traditionally, manual rammers have been used for compaction but nowadays pneumatically powered dynamic rammers are commonly used. Once the soil has been adequately compacted the formwork is removed, often immediately after compaction, leaving the finished wall to dry out. Walls are typically 300–450 mm thick, but this can vary widely according to design requirements.

Rammed earth walls often exhibit a distinctive layered appearance as a result of the construction process, corresponding to the successive layers of soil compacted within the formwork (Figure 3). This attractive appearance is

(Grimshaw architects; In Situ Rammed Earth; 1999)
Figure 1 Rammed earth wall construction at the Eden Project, Cornwall

Figure 2 Construction of a rammed earth wall

(Architect: Sassenroth & Reitermann; Martin Rauch; 2000)

Figure 3 Rammed earth wall finish, Chapel of Reconciliation, Berlin

undoubtedly one of the appeals of rammed earth construction and as a result walls are often left without plaster or render.

1.3 Brief history and development

Both loadbearing and non-loadbearing walls in a wide variety of structures and applications have been constructed using the rammed earth technique over many centuries[4]. Rammed earth walls are found throughout the world, with many examples throughout Asia, including sections of the Great Wall of China, in Africa, Latin America and Europe (Figures 4–7).

Rammed earth was probably introduced to Britain by the Romans and so has been used in the UK for around 2000 years[5]. However, the most significant period of construction followed its reintroduction into the UK in the early nineteenth century, following a revival of interest in France led by

Figure 4 Traditional rammed earth building, Morocco

Figure 5 Seven-storey rammed earth building, Weilburg, Germany (c1820)

Figure 6 Rammed earth building, Rhone Valley, France

Figure 7 Rammed earth walling at the Alhambra, Granada, Spain

Brief history and development

Figure 8 Victorian five-storey rammed chalk houses, Winchester, Hampshire (c1840)

Figure 9 Victorian rammed chalk building, Andover, Hampshire

Figure 10 Rammed chalk house, Amesbury, Wiltshire (c1920)

François Cointeraux[6]. Throughout the nineteenth century a number of rammed earth and rammed chalk buildings were erected in southern England. Examples include five-storey loadbearing rammed chalk houses in Winchester built around 1840 (Figure 8) and numerous country houses built around the same time (Figure 9). After the First World War a small number of experimental rammed earth and chalk houses were constructed in Amesbury, Wiltshire, many of which are still in use today (Figure 10)[7].

In recent years an increasing number of rammed earth and rammed chalk projects have been completed in the UK. Applications to date include the Eden Project Visitors Centre (Figure 11), Centre for Alternative Technology's AtEIC Building (Figure 12), a Chelsea Flower Show prize winning exhibit wall (Figure 13), sections of Woodley Park Sports Centre (Figure 14), the Kindersley Centre (Figure 15), Bird-in-Bush Nursery development (Figure 16) and Mount Pleasant Ecological Business Park (Figure 17). The stimulus for this development has primarily been the desire to reduce the environmental impact of building and to explore more sustainable and natural building methods. However, recognition of the aesthetic qualities of rammed earth construction has also been a significant factor in its re-emergence, especially in works by Martin Rauch[8] and Jörg Depta (Figures 18–20).

(Grimshaw architects; In Situ Rammed Earth; 1999)
Figure 11 Eden Project Visitors Centre

(Pat Borer; Simmonds-Mills; 2000)
Figure 12 AtEIC Building, Centre for Alternative Technology, Machynlleth, Powys

(David Clark)
Figure 13 Wall at Chelsea Flower Show 2000

(John Renwick; 1999)
Figure 14 Woodley Park Sports Centre, Skelmersdale, Lancashire

Brief history and development

(Alec French & Partners Architects; Mark Lovell Design Engineers; 2003)
Figure 15 Rammed chalk walls, Kindersley Centre, Sheepdrove Estate, Berkshire

(JM Architects; In Situ Rammed Earth; 2004)
Figure 16 Bird-in-Bush Nursery, London

(Tim Stirrup & John Wray; 2004)
Figure 17 Mount Pleasant Ecological Business Park, Porthtowan, Cornwall

(Martin Rauch; 2000)

Figure 18 Altar, Chapel of Reconciliation, Berlin

(Architect Guenther Ludewig; LehmBauWerk, Berlin)

Figure 19 Rammed earth wall, Brandenburg, Germany

(LehmBauWerk, Berlin)

Figure 20 Rammed earth wall, Zeesen, Germany

Brief history and development

Stabilised rammed earth is a specific form of rammed earth construction that uses sub-soils combined with stabilising agents to improve the material's physical characteristics. Ordinary Portland cement is by far the most common additive used. Applications of stabilised rammed earth have been varied. Over the past 30 years stabilised rammed earth has been used increasingly in Australia and the USA (Figure 21), with a few thousand buildings now built. In some areas of Western Australia, such as Margaret River, stabilised rammed earth represents a significant proportion of new building work (Figure 22). Recent applications of stabilised rammed earth in the UK include two private houses (Figures 23 and 24).

Figure 21 Stabilised rammed earth house, Rural Studio, Alabama, USA

Figure 22 Stabilised rammed earth house, Western Australia

(David Sheppard Architects, Cob Construction Co; 1997)
Figure 23 Dragons Retreat, Devon (stabilised rammed earth)

(Tim Hewitt; 2003)
Figure 24 Jasmine Cottage, Norfolk (stabilised rammed earth)

1.4 Advantages and limitations of rammed earth

1.4.1 Benefits and qualities

There are many reasons for choosing to use rammed earth construction. Some of the most significant of these are as follows.

- **Sustainable construction**

  As a natural material, without processed additives, rammed earth can have significantly lower embodied carbon dioxide and energy than conventional manufactured building materials, as well as reduced toxic chemical content and fewer emissions from industrial processes. Sub-soils, without binder stabilisation, can also be readily re-used, recycled or disposed of without risk of contamination to the environment. Though strictly speaking the raw materials for rammed earth are non-replenishable, they may in a wider context be considered effectively inexhaustible.

The main environmental impacts of rammed earth stem from transportation of heavy materials. Using in-situ sourced materials is the ideal but is entirely dependent on the material's suitability. Raw materials are therefore often brought to the construction site from external sources. There is some potential for using quarry and other construction spoil materials in rammed earth. Suitable materials are likely to be found within an acceptable distance of most if not all areas of the UK.

- **Architectural quality and flexibility**

Rammed earth can be a particularly beautiful finished building material, sometimes likened to sedimentary rock deposits, and offers a very wide variety of textures and colour finishes (Figures 1–20). Characteristically, rammed earth is built-up in a series of approximately horizontal layers of compact soil between 50 and 100 mm deep. Each layer corresponds to the incremental construction sequence. The appearance of each compaction layer varies, with the uppermost material closest to the compactor head denser than the material below (Figure 25). This variation in material density highlights the overall layered or stratified finish.

A variety of textural and colour finishes can be achieved through careful selection and preparation of materials. Materials with comparatively high fines content, often placed slightly wetter than optimum, will tend to lose their stratified appearance. The stratified appearance can be exaggerated by introducing varying layers of colour (Figures 19 and 20), tooling the surface finish (Figure 26) and use of other materials (Figure 13). Walls may be vertical or battered in section (Figure 1), or linear or curved in plan (Figure 19), and wide varieties of finishes and forms are possible, only really limited by constraints of the shuttering and compaction processes.

Rammed earth building is suited to a variety of small- and large-scale building projects. Earth is also a natural and universal building material, providing pleasant living spaces. Many rammed earth building operations can be undertaken by a relatively inexperienced labour force, though efficient management is needed. As a labour-intensive technology, earth building is well suited to community-based low-cost housing schemes in both economically developed and developing countries, as well as to the broader owner-builder market.

Figure 25 Compaction layers in rammed earth

Figure 26 Tooled finish in rammed earth

- **Contribution to building health and performance**
  Clays within rammed earth soils are hygroscopic, releasing or absorbing moisture in response to changing local atmospheric conditions. Studies[9] have proven that earth walls are very effective in regulating the internal relative humidity to between 40 and 60%. This property of unstabilised earth walls reduces stress on the building fabric and improves indoor air

quality, removing asthma triggers and reducing respiratory diseases.

As a dense and bulky material, rammed earth has considerable thermal mass, and has the ability, together with other appropriate design, to contribute to passive environmental performance of the building.

- **Improved performance of earthen materials**
  In comparison with other forms of earthen construction, such as cob and adobe, rammed earth construction has higher strength and stiffness, reduced drying shrinkage arising from compaction at relatively low moisture content, improved durability and greater density.

- **Speed of wall construction**
  Rate of wall construction is typically between 5 and 10 m^2/day for a 300 mm thick rammed earth wall for a team of three to four workers. After removal of the formwork, walls should require little further attention and so the overall speed of wall construction can be very favourable in comparison with finished concrete block, brick and stone masonry wall construction.

1.4.2 Limitations and drawbacks

Rammed earth is certainly not a universal panacea for sustainable building. Whilst there are many situations where rammed earth is entirely suitable, its limitations, weaknesses and drawbacks need to be carefully considered during selection and design. Though likely to vary with project details, the most significant issues for consideration are set out here.

- **Durability of rammed earth**
  Rammed earth is susceptible to decay in the presence of water. This requires special consideration in design and construction and throughout its service life. New walls in particular should be protected from inclement weather to prevent premature damage. Generally the level of maintenance required will be higher than that of some competing materials, but it is also inversely related to the extent of protection to walls provided by roof and base details.

  Strength and durability concerns limit the building form. Wall construction is limited to sites where there is little risk of flooding, although walls in sites prone to flooding may be built on extended plinths above likely flood water levels. External walls are generally raised on upstands and often protected with large eaves extensions. Corners and other fragile edges require protection and should generally be chamfered. Applied protective coatings can be used to avoid excessive 'dusting' (loss of friable surface material) from wear and tear on internal surfaces. However, it is generally preferable to leave walls untreated as coatings have proven problematic and they can impair other material characteristics, such as hygroscopic performance.

Rauch[10] and others have recently proposed that exposed rammed earth walls should be allowed to weather naturally without the protection of protective coatings or additives. This approach is not new of course; vernacular earthen buildings are routinely repaired after each rainy season around the world. To minimise the risk of premature failure requires confidence and prior knowledge of material performance under the expected weathering conditions.

Durability concerns make rammed earth walls particularly well suited to internal, protected, applications as part of other building structures, such as timber frame construction (Figures 3, 12, 15, 19 and 20). Walls contribute to structure (though may also be non-loadbearing), environmental (thermal and humidity) regulation, and provide both acoustic and fire barriers.

- **Wall thickness**

To ensure sufficient lateral resistance and allow construction, rammed earth walls are normally 300 mm or more thick, typically thicker than many other forms of modern construction. Wall thickness is governed by low material strength and compaction requirements. Low material strength (Chapter 3) also places restrictions on the size and spacing of openings. Slenderness (height-to-thickness) ratio is often limited to 12 for loadbearing walls (Chapter 6).

- **Thermal resistance**

Rammed earth derives much of its physical resistance from the material's relatively high density, but a consequence of this is its poor thermal resistance. To meet thermal performance levels expected of modern energy efficient buildings and to meet requirements of the Building Regulations, external rammed earth walls must either be very thick (>700 mm) or use additional insulation materials. External insulation combined with rain-screen cladding can be used to improve weathering resistance.

The poor insulating qualities of rammed earth might be accepted without modification, and in a more holistic approach other elements (other external walls, plinths, floor, roof) can be 'super-insulated', and together with fuel efficient heating measures ensure an overall satisfactory building performance.

- **Material selection and variability**

Not all soils are suitable for rammed earth construction. Only sub-soils, rather than top soils, should be used. Soil should be well graded between gravel and clay-sized particles. Poor-quality sub-soils can be improved for rammed earth by mixing them with missing soil fraction(s), a process sometimes referred to as granular stabilisation.

Natural soils can have a very wide range of properties so they must be carefully selected for each project. Extensive laboratory testing is often needed to ensure appropriate performance, and advice from a soils specialist must sometimes be sought. To improve consistency and reliability, and remove the cost and uncertainty of material testing with each new project, many established rammed earth builders repeatedly take materials from quarry sources, sometimes resulting in transportation of materials over considerable distances. Costs are also increased if materials have to be imported.

- **In-situ construction**

Rammed earth is primarily built as an in-situ shuttered form of construction, which places particular demands on both design and construction. Wall design must allow shuttering to be erected and dismantled repeatedly during construction. Wall layout should therefore be co-ordinated with the modular nature of the adopted formwork system. Rammed earth works require space for storage of materials and equipment and the movement of plant. The shuttering and partly built walls must be protected from adverse weather and other construction operations. Materials should be placed at their optimum moisture content and so soil must be stored and prepared accordingly. Quality control of the in-situ compaction of materials inside shuttering can be problematic. The higher-density rammed earth walls, compared with other earth-building techniques such as cob, require more sub-soil material. Some of the problems of in-situ construction are being addressed through prefabricated rammed earth (Section 2.1).

1.4.3 Economic cost

The finished cost of rammed earth construction varies greatly, depending on the specifications and requirements of the wall finish. As with other materials, highly sculptural work will be more expensive than more simple functional rammed earth walling. Experience has proven that the cost of general quality rammed earth can be comparable to or even cheaper than alternative forms of fully finished masonry wall construction[11]. At the time of publication the finished cost of general quality rammed earth was approximately £70–100/m^2 for a 300 mm thick wall.

Though the raw materials are relatively inexpensive, labour costs associated with the handling of materials and shuttering comprise the main cost of construction. Formwork systems must therefore be used efficiently and materials preparation well planned and controlled. Handling of formwork typically accounts for 25–50% of construction time, and so simplifications in the formwork scheme can provide significant cost savings. Walls curved in plan or battered in section will generally be more expensive than linear walls. Labour costs can of course be reduced or eliminated altogether through volunteer labour or a self-building approach, as has been practised successfully in a small number of recent projects, such as the Woodley Park Sports Centre (Figure 14).

1.5 Structure of the guidelines

Chapter 2, *Preliminary design considerations*, sets out guidance on applications and uses of rammed earth construction, how rammed earth influences other design and planning decisions, how rammed earth is influenced by the provisions of the Building Regulations, and briefly some initial contractual considerations. Chapter 2 is primarily intended for those unfamiliar with rammed earth.

Materials for rammed earth construction, Chapter 3, gives advice on the selection of materials for rammed earth and typical physical characteristics of rammed earth materials. Designers and contractors in particular should refer to this chapter. Guidance on test procedures for rammed earth material is set out in detail in Appendix A and a specification for rammed earth is in Appendix B.

Chapter 4, *Construction of rammed earth walls*, is a practical guide to the process of wall construction. It should appeal to contractors, designers and anyone else seeking greater understanding of the practical aspects involved in construction. Various details, including base, eaves, insulation and protective coatings, are outlined in Chapter 5, *Details for rammed earth construction*. The chapter is primarily intended for designers, especially architects, and building control officers.

Structural engineering design of walls is set out in Chapter 6 together with more detailed provisions in Appendix C. Provision for maintenance and repair is given in Chapter 7, including advice on defects in new building as well as methods for protecting and repairing older walls.

Appendix D briefly discusses the merits and disadvantages of stabilised rammed earth construction together with advice for material selection and physical properties.

Finally, details of individuals and organisations with varying experience and expertise in rammed earth construction are given, followed by a glossary, reference list and bibliography.

2 Preliminary design considerations

This chapter describes the various applications and uses of rammed earth construction, including walls, roofs and floors, as well as more novel uses such as sculpture and furniture. The influence of rammed earth construction on related design and construction aspects is outlined. The impact of Building Regulation requirements on rammed earth construction is also discussed.

2.1 Applications

Rammed earth has been used in a variety of building applications, including loadbearing and non-loadbearing walls in buildings, free-standing internal and external walls, retaining walls, cladding panels facing other forms of construction, flooring, and even for furniture and in utilities such as fireplaces. Walls in buildings may be external, providing thermal and acoustic insulation as well as shelter. Internal walls must provide acoustic insulation as well as thermal mass. Each potential application is discussed below.

2.1.1 External walls

External walls in buildings may be either loadbearing or non-loadbearing. In both cases walls must provide adequate thermal and acoustic protection from the external environment. The acoustic performance of thick, dense walls is generally good.

Thermal conductivity of rammed earth is relatively high (around 0.8–1.5 W/mK) and so solid walls 300–500 mm thick will not provide adequate thermal insulation to meet modern building requirements. The options available to improve thermal insulation are listed below. In all cases at least one face of the wall is obscured by the insulation layer.

- **External lime render with lightweight aggregates:** render coat up to 100 mm thick enhances thermal insulation and provides weathering protection. A 300 mm thick solid rammed earth wall with lightweight aggregate render can achieve U-values in the range of 0.6–0.7 W/m²K.
- **External insulation with rain-screen cladding or render:** thermal insulation value can be selected to meet requirements whilst maintaining the thermal mass benefit internally, and screening provides weathering protection.

- **Internal insulation and stud walling:**
  the external appearance of the rammed earth building is maintained, though the thermal mass of the walls is isolated from the internal space by the insulation. Stud walling and boarding also protects walls from the risk of internal abrasive damage, including dusting.

Details for thermal insulation are outlined in Section 5.8. The addition of lightweight materials such as vermiculite, pumice, natural fibres, polystyrene waste and cork to reduce thermal conductivity of rammed earth, has largely proved unsatisfactory partly owing to the high pressures exerted during compaction.

2.1.2 Internal walls

Internal walls may also be loadbearing or non-loadbearing. The long-term movement from shrinkage and possible creep of loadbearing walls, if used with other structural elements, such as timber frame, require consideration. Non-loadbearing walls inside frame construction may need to be checked for robustness and stability, but can contribute to racking resistance. Internal walls can also provide good acoustic separation.

Though internal walls should be fully protected from weathering once the roof is in place, internal surfaces may be subject to abrasion through general use or deliberate vandalism. Raised plinths, protective coatings, screens and barriers can be used to protect wall surfaces. Internal walls also provide effective thermal mass and humidity control in otherwise lightweight building forms, such as timber frame.

2.1.3 Free-standing walls

Free-standing walls may be used externally as boundary walls or internally as dividing or parapet walls. Internal free-standing walls can also be used for heat storage (thermal mass) and humidity control in buildings largely built of other materials. External rammed earth walls generally require further protection against weathering, such as a coping and render coat.

2.1.4 Retaining walls

Though not generally recommended, because of persistent damp conditions and weathering exposure, rammed earth has been used historically for retaining walls or facing walls to the sides of structurally stable cuttings. In modern applications stabilised rammed earth has been used for retaining walls in relatively mild frost-free environments.

Applications

2.1.5 Pre-formed rammed earth

In recent years, in line with the general move towards off-site fabrication of building elements, pre-formed or prefabricated rammed earth has developed. To date, prefabrication has been used by only a very small number of specialist overseas practitioners[8], and the wider use of pre-formed rammed earth is largely unproven in the UK. Prefabrication potentially allows higher-quality factory construction of elements under sheltered conditions whilst also minimising on-site construction time. Examples to date include large wall blocks (Figure 27) as well as 100–200 mm thick cladding panels. Although costs are likely to increase, owing to transportation and lifting requirements, the use of prefabricated rammed earth is likely to increase in forthcoming years.

Figure 27 Prefabricated rammed earth walls

2.1.6 Rammed earth flooring

Earth and chalk floors have a long and unbroken vernacular tradition (Figures 28 and 29). Rammed earth floors are typically compacted in single or multiple layers of 100–300 mm total depth. Lower layers are composed of coarser material, with top layers using well graded material smaller than 8 mm. Floors are typically laid on prepared ground or hardcore, sometimes even

Photo: Martin Rauch

Figure 28 Rammed earth floor

Photo: John Wray

(Tim Stirrup & John Wray; 2004)

Figure 29 Rammed earth floor, Mount Pleasant Ecological Park, Porthtowan, Cornwall

without a damp-proof membrane, which allows water vapour to evaporate through the floor. In general, however, damp-proofing measures are advisable and indeed a Building Regulation requirement. As the floor dries, efflorescence (salt deposits) may occur on the surface, though it is usually readily removed by brushing. Chalk floors in East Anglia are typically finished off with a lime and sand screed.

To improve wearing and water resistance, floors are often stabilised with oils, such as oxidised (boiled) linseed oil, and waxes, including carnauba and beeswax. Underfloor heating systems may also be incorporated into rammed earth floors. Rammed earth floors are regularly re-oiled or waxed during cleaning, and routine maintenance can provide remarkably durable surfaces. However, the use of protective coatings such as linseed oil may lead to excessive atmospheric emissions of volatile organic and other compounds.

2.1.7 Other uses

In recent years rammed earth has been used for a variety of innovative items of furniture, sculpture and utilities, including fireplaces, stoves and church altars (Figures 18 and 30). The material is generally much finer than in wall construction and often contains additives such as natural resins, oils or waxes.

(In Situ Rammed Earth; 2003)

Figure 30 Office desk, Engineers HRW office, London

2.2 Influence of rammed earth on other construction activities

2.2.1 Shuttered form of construction

As a shuttered in-situ form of wall construction, rammed earth operations are likely to influence other construction activities and the general scheduling of construction works. The impact and consequence of these may be minimised by careful design. Wall layout and sizes should reflect the modular nature of the formwork system adopted. Access for erection and dismantling of the formwork is required. Wall construction is often a critical path activity, controlling the rate of progress during that phase of the project. One of the most significant considerations includes the need to provide protection and support whilst also allowing the walls to dry out.

2.2.2 Organisation of site

In-situ rammed earth requires space, and time, for the storage and preparation of materials, and for the storage, erection, cleaning and dismantling of formwork shuttering. This should be considered throughout planning and design as it will govern the rate of wall construction. After removal of formwork, walls may need temporary propping during drying. Space for props and their effect on other works therefore need to be considered.

2.2.3 Productivity

The productivity of rammed earth construction is dependent on many related factors, some easily controlled, such as formwork type, site organisation, height of working and method of compaction, and others that are not, such as the weather. Formwork handling typically accounts for some 25–50% of site work. Productivity rates for finished wall construction vary, but for a typical 300 mm thick wall, output is in the range of 5–15 m^2 per day for an experienced team.

2.2.4 Weather conditions

Weather is likely to have an important influence on the timing and rate of construction. Inclement weather, rain and high winds will prevent rammed earth work unless it is carried out under cover. The risk and consequences of losing days to inclement weather therefore need to be considered in the project planning. Working under cover therefore has significant advantages, though sufficient space is required beneath the roof for compaction (Figure 31). Stockpiled materials also need covering to prevent over-wetting from rainfall.

Influence of rammed earth on other construction activities

Figure 31 Rammed earth wall construction under cover, Centre for Alternative Technology

2.2.5 Protection

Rammed earth walls require protection from weathering and other construction activities. Ideally roofing protection should be available as soon as possible following construction. Non-loadbearing walls may be built inside or under roofing. Alternatively loadbearing walls may also be built under cover using temporary props for the roof structure. Other construction works should be planned and carried out to avoid damage to wall surfaces. Where necessary, walls should be protected by barriers or covers. Corners and edges need protection in particular.

Covering of walls with plastic sheeting or screens can provide protection where roofing is not available. Care is required to ensure that covers remain intact as gaps can channel concentrated flows of rainwater, leading to significant localised damage. All covers should therefore be regularly inspected. Tightly wrapped plastic covers will reduce the rate of drying of walls, further delaying general rate of project progress.

2.2.6 Services

Ducting for electrical services can readily be rammed into walls. Block-outs provide space for sockets and switches. Therefore careful planning and consideration is required for placement and general provision. Round ducts are easiest for placement and compaction of materials. Details for services are covered in Section 5.6.

2.2.7 Material shrinkage

Where rammed earth walls are used as loadbearing members in combination with other elements, the shrinkage and rate of drying need to be carefully considered in design and accommodated in construction (Section 3.6.4). Where differential shrinkage is expected to be significant, temporary propping may be required for some months as rammed earth slowly dries out. Alternatively, shrinkage can be taken into account using folding wedges or jacks. Shrinkage may require monitoring in the period following construction. Where renders or other protective coatings are to be applied, further wall shrinkage needs also to be considered.

2.2.8 Health and safety

Health and safety issues of rammed earth construction include working at height (formwork activities and compaction), lifting formwork and props, noise of pneumatic compaction, the stockpiling of materials, dust from dry materials, use of vibratory equipment (risk of vibration white finger), and the use of heavy lifting plant. As well as the direct impact of these on rammed earth contractors, the health and safety impact on others on or adjacent to the site, including the public (noise, dust and vibration), needs to be considered.

2.3 Building control

The Building Regulations ensure the health and safety of people in and around buildings by providing functional requirements for building design and construction in England and Wales. A separate system of control applies in Scotland and Northern Ireland. The Building Regulations also promote energy efficiency in buildings and contribute to meeting the needs of disabled people through requirements for accessibility. Compliance with the Building Regulations is a legal requirement for all new building works. Demonstration of compliance is most commonly through building control officers appointed by the local authority, though independent approved inspectors are also able to undertake this work.

Recent experience across a number of local authorities in the UK, demonstrates that rammed earth and stabilised rammed earth walls are able to satisfy the requirements of modern building regulations. Failure to meet requirements of building control officers has rarely been cited as a reason for the failure of proposed rammed earth projects. The most significant Building Regulation[12] requirements with respect to rammed earth construction are summarised below.

2.3.1 Part A – Structure

Part A of the Building Regulations (2004 edition) requires the building to be constructed so that the combined loads and ground movements can be resisted safely and without undue deflection or deformation. Rammed earth has proven suitable for both loadbearing and non-loadbearing construction in a variety of building types and situations. Accidental damage to loadbearing walls due to flooding, including burst water pipes, needs to be considered in design.

The influence of increasing moisture content on material strength, potential loss of section through erosion, the rate of drying and associated rate of strength development, the amount of drying shrinkage and influence of defects (poor compaction) on wall capacity are important considerations for design and construction. Material characteristics (compressive strength and drying shrinkage) should generally be assessed by initial testing and checked during construction. Test procedures for rammed earth materials are outlined in Appendix A.

2.3.2 Part B – Fire safety

Part B of the Building Regulations (2000 edition) requires the building to be designed and constructed to ensure safe and effective means of escape from fire. In accordance with Table A6 of Part B rammed earth can be defined as a non-combustible material as it very rarely contains more than 1% (by weight or volume) of organic material. A 300 mm solid rammed earth wall will provide fire resistance of at least 90 minutes. Movement joints in separating walls should be located with care and sealed with foam strips or similar with intumescent properties to minimise any interruption to the fire barrier. Loadbearing rammed earth walls might be damaged by high-pressure water hoses; the potential accidental damage and loss of material should be considered in design. Movement joint details are discussed further in Sections 4.2.7 and 5.9.

2.3.3 Part C – Resistance to moisture

Part C of the Building Regulations (2004 edition) requires walls to provide adequate protection to the building and its occupants from the harmful effects caused by ground moisture, precipitation and wind-driven spray and condensation. The installation of damp-proof courses and membranes in base details prevents rising damp (Section 5.2). Penetration of moisture from rain and snow is limited through the absorption and dissipation of the moisture throughout the fabric of the solid rammed earth wall. In drier periods this moisture is released to the atmosphere as part of a breathing wall system. In general, the level of exposure to wind-driven rain is greatly reduced compared with other solid walls owing to the desire to protect walls with extended eaves, raised plinths and other means to minimise weathering erosion. Vapour-permeable protective coatings can also reduce the amount of rainwater directly entering a rammed earth wall (Section 5.5). However, water-sealant protective coatings to prevent moisture entering rammed earth walls are not recommended as they can also trap moisture inside the wall fabric.

2.3.4 Part E – Resistance to the passage of sound

Part E of the Building Regulations (2003) requires that walls provide protection against sound transmission from other parts of the building and adjoining buildings. Dense solid rammed earth walls provide very effective acoustic separation (Section 5.9). Where floors are directly supported by separating rammed earth walls, due consideration is required to minimise direct sound transmission. In this regard rammed earth walls may be considered as equivalent to other solid masonry walls, and floors may be detailed accordingly. Movement joints in separating walls should be located with care and sealed with expansive strips to minimise any interruption to the sound barrier (Section 5.9).

2.3.5 Part L – Conservation of fuel and power

Part L of the Building Regulations (2002) requires that reasonable provision be made to limit heat loss through the building fabric. Compliance can be achieved in three ways: the elemental method, the target U-value method and the carbon index method. The carbon index method is widely considered to be the most appropriate and easiest way of demonstrating compliance with Part L for earthen wall construction.

The thermal transmittance (U-value) of a 300 mm thick solid rammed earth wall is between 1.5 and 3 W/m^2K. These values are considerably higher than either of the maximum U-values of 0.35 W/m^2K required in Part L (2002) and 0.7 W/m^2K required in Approved Document C (2004)[13] to control condensation. To comply with these requirements external rammed earth walls therefore need additional insulating finishes or layers. Insulation details are set out in Section 5.8.

Part L of the Building Regulations is due for revision in 2005. Proposals currently out for public consultation show the elemental, target U-value and carbon index methods being replaced by a single method based on annual carbon emissions calculated according to a revised version of the Standard Assessment Procedure (SAP)[14].

2.3.6 Regulation 7 – Materials and workmanship

Regulation 7 of the Building Regulations (2000) requires that building work be carried out with adequate and proper materials and in a workmanlike manner. Fitness of rammed earth materials will generally be established by sampling, laboratory testing of materials or precedence. In-situ testing of rammed earth is difficult and therefore material compliance testing is normally undertaken on specimens (cylinders or cubes) prepared for this purpose. Adequacy of rammed earth quality of work is measured against the provisions of the specification, test panel(s) and previous works.

2.4 Contractual considerations

Contracts for rammed earth construction may be let to specialist subcontractors or general building contractors under standard forms of building contract. The finest quality work will generally require specialist contractors experienced in rammed earth construction. Contractual responsibilities need to be clearly defined in respect of material selection, material compliance testing and protection of walls after construction. Protection from weather and accidental damage during works is essential and responsibility must be clearly stated. Specifications for rammed earth works may follow those outlined in Appendix B. For quality control purposes test panel(s) should be constructed on site to allow inspection, and material specimens for compliance testing should be agreed in advance.

3 Materials for rammed earth construction

This chapter describes raw materials used in rammed earth construction and sets out recommendations for their selection. The role of compaction is outlined, followed by a discussion of the physical characteristics of rammed earth, including strength, drying shrinkage and durability.

3.1 Raw materials

3.1.1 Soil

The primary constituent of rammed earth is inorganic sub-soil, taken from deposits found beneath the organic and life-sustaining topsoil. The physical and chemical characteristics of sub-soils, which govern their suitability for rammed earth, depend primarily on the original parent rock geology and subsequent weathering, including hydrological and hydro-geological processes, and possible further changes on exposure to the atmosphere. Sub-soil layers may change type, colour and texture with depth as a consequence of the formation processes they have undergone. In addition to natural deposits, soils produced as by-products of other mineral industries, such as stone processing, might also prove suitable as constituents of rammed earth.

Sub-soil structure may be considered as comprising four main particle types, classified according to their size. The grading, or relative proportions of *gravel, sand, silt* and *clay* govern the properties and suitability of soil for rammed earth building. All elements play an important role in performance of the material. Gravel provides the inert skeleton or matrix, together with the sand fraction, and enhances weathering resistance of exposed faces. Rammed earth relies on its clay fraction to bind the other largely inert particles together. Clays, formed during chemical weathering, are quite different from other particles, swelling when wetted and shrinking as they dry out. Sensitivity of the clay fraction to variations in moisture content depends on mineralogy and has a very important influence on soil properties. The main clay mineral types include kaolinite (the least active), illite, and smectites (the most expansive and least common), which include montmorillonite.

3.1.2 Water

The quantity and quality of water present has a very important influence on the overall quality of rammed earth. Water sourced from other than the mains supply should not contain excess organic matter, soluble salts or other materials likely to prove deleterious. Most important to the quality of construction is the quantity of water present at the time of compaction. Rammed earth should be placed at the material's optimum moisture content (Section 3.3).

3.1.3 Other constituents

Air, together with free water, is present in the interstitial voids between soil particles. Compaction seeks to minimise voids ratio and hence the air content of rammed earth.

Soluble salts, water-soluble chlorides, sulfates and carbonates occur naturally in soils. As rammed earth dries out, unsightly salt deposits may appear on the surface (efflorescence). These normally can be cleaned off when drying is complete, but sometimes efflorescence can lead to surface damage where salt deposition is heavy or hindered at the surface. Soluble salts can also be detrimental to the efficacy of some additives.

Soils with high organic matter content should generally be avoided in rammed earth. Organic matter may subsequently decay and rot, and lead to deterioration of the wall fabric. It also increases susceptibility of material to insect attack, and is detrimental to the action of some additives.

Concern regarding the presence of harmful contaminants in sub-soils has been raised during a small number of recent projects. Though arsenic occurs naturally in soil the levels are not considered to be a significant widespread problem in the UK, especially in sub-soils. However, in some areas, such as Devon and Cornwall, which have been subject to significant mine workings in the past, the levels can be significantly higher than recommended concentrations of 10 to 20 mg of toxic carcinogen per kilogram of soil. In areas where arsenic concentration levels have been identified as a concern, testing contaminant levels may be prudent during initial analysis of soils, though the general risk is considered very low for rammed earth. The British Geological Survey has recently introduced the Physiologically Based Extraction Test (PBET) to determine the amount of dangerous fraction of arsenic available rather than the total quantity in a soil. A number of specialist laboratories nationwide are able to undertake chemical analysis of soils.

3.1.4 Rammed chalk

Rammed chalk construction is a particular type of wall construction using the rammed earth technique found in some regions of Britain, such as central southern England, where suitable deposits of chalk are readily available[6]. Examples are also found in Sweden[15]. The method of construction differs little from rammed earth; walls are formed from chalk rubble rather than from clay-bearing sub-soil. The excavated chalk is broken down into fragments before ramming.

Chalk is a sedimentary rock of between 70 million and 100 million years old. It was formed from coccoliths (lime secretions of algae) and may contain foraminifera and other larger calcareous shells. Chalk has characteristics quite distinct from the clay-bearing sub-soils used for rammed earth. Typically the plasticity index is much lower, with plastic and liquid limits of chalk around 21% and 27% respectively. The density of rammed chalk is much lower than that of natural chalk, varying between 1300 and 1720 kg/m^3. Though there is little published information, compressive strengths of rammed chalk walls will normally[6] attain at least 0.3–0.5 N/mm^2.

3.2 Soil characteristics

A range of soil characteristics influences selection of a material for rammed earth construction. Testing of soils in preparation for rammed earth works may be undertaken in accordance with procedures developed for more general civil engineering works, as currently set out in BS 1377-2[16].

(a) Colour

Natural soil is available in a very wide range of colours, including reds, yellows, browns, greens, blues, white and grey. Colour is indicative of constituents but is also a very important consideration in soil selection for rammed earth construction. Black and dark brown soils often have high organic matter content. Red, reddish-brown, yellow, and yellowish-brown are indicative of the presence of iron. Variation in aggregate colour can lead to non-uniform finishes. Though other parameters, such as strength and erosion resistance, are more likely to govern soil selection, colour is an important aesthetic consideration for the client and designer. Natural colours can be varied by using binders, pigments or by blending different soils. The use of varying coloured soils has been used very effectively by a number of builders to enhance the stratified (layered) finish (Figure 32). Some protective coatings can alter colour, sometimes favourably, and so should be checked before use.

(b) Grading

Grading of solid particles by size is widely used to describe composition and suitability of soil for rammed earth

(Lehmbauwerk, Berlin)

Figure 32 Compaction layers in rammed earth

construction. Soil grading is normally determined by sieving and sedimentation testing (BS 1377-2)[16]. Recommendations for clay, sand and gravel content are outlined in Section 3.5.3. Grading will also have a significant influence on finished texture and surface integrity (friability) of rammed earth. However, the influence of variation in grading on physical characteristics of rammed earth, including strength and durability, remains uncertain owing to lack of test data.

(c) **Density**
The densification of soil, by expulsion of air voids through compaction, is intrinsic to rammed earth construction. Improving density generally improves strength and durability, though it also increases thermal conductivity. Soil density depends on moisture content, compactive effort, composition and grading. The dry density of poorly graded soils can also be increased by the addition of particle sizes lacking in the original matrix.

(d) **Plasticity**
Plasticity is the ability of soil to undergo non-recoverable deformation at constant volume without crushing or cracking. The property is primarily due to the presence of clay minerals. Soil plasticity may be described in terms of liquid and plastic limits, moisture contents that mark the transition between liquid and plastic states, and plastic and solid states, respectively. Soil

plasticity index, the moisture content range over which soil acts plastically, is the difference between the liquid and plastic limits. Other important characteristics for rammed earth, such as drying shrinkage, cohesion and rate of drying, are related to soil plasticity. These are determined in accordance with procedures set out in BS 1377-2[16].

3.3 Soil compaction

Compaction improves strength, dimensional stability and durability whilst reducing permeability, porosity, and compressibility. Compaction in rammed earth construction is generally dynamic, through the use of pneumatic (Figure 33) or manual rammers (Figure 34), though vibratory compaction, using wacker plates, has been used in sufficiently thick walls and, more commonly, floors.

For any given soil mix and level of compactive effort (energy) there is an optimum moisture content at compaction that achieves a maximum dry density (Figure 35). The optimum moisture content reduces as the compactive effort increases. The optimum moisture content and corresponding maximum dry density are commonly determined in accordance with the heavy manual compaction test[17] (Appendix A.3.2). Optimum moisture content and density may also change with the use of additives.

Figure 33 Pneumatic rammer

Figure 34 Manual rammer

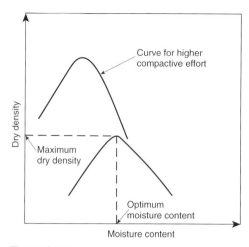

Figure 35 Relationship between compaction moisture and dry density

3.4 Additives

A range of additives may be used to improve strength and water resistance or reduce drying shrinkage. In this section natural fibre additives to rammed earth are discussed. Binders and pozzolanic additives, which form stabilised rammed earth, are discussed further in Appendix D. Materials used primarily for surface protection are considered in Section 5.5.

Natural fibres, such as straw, are widely used in mud construction, including adobe and cob, as a means of controlling drying shrinkage and improving tensile strength. The inclusion of fibres also reduces density, which indirectly improves thermal resistance. Straws from all the main cereals (wheat, barley, rye, oats and maize) are considered suitable, though wheat is generally preferred. Other suitable smaller-diameter fibres include flax, hemp and sisal.

As drying shrinkage is typically much less in rammed earth than in cob and other forms of wet earth construction, natural fibres have traditionally not been added. However, in some recent rammed earth projects where soil shrinkage has been a concern, small fibres have been added to the mix. For example, the soil mix at the Chapel of Reconciliation in Berlin included a small quantity of flax fibres.

Furthermore excessive use of fibres is likely to be detrimental as it inhibits binding with the soil and, for best results, fibres must be well mixed throughout the soil. Fibre content in cob and other mud construction is typically 1–2% by mass, though the content in rammed earth is likely to be much lower. Using fibres, and other additives, increases the complexity of soil preparation as mixing is required. Mixing fibres uniformly throughout the mix can be more difficult owing to aggregation and balling. Pneumatic soil compaction may damage larger fibres.

3.5 Soil selection

Selection of an appropriate raw material is critical to the success of rammed earth. In-situ-sourced materials often prove suitable though they may require modification, such as blending with other materials. This is particularly true as rammed earth buildings are built in areas of the UK without a significant vernacular tradition of earth building. Increasingly, established rammed earth contractors use materials from a quarry and import materials to site. This reduces the costs of material testing and minimises risks associated with untried materials, but increases the environmental impact through transportation.

Though quarry over-burden material might prove a suitable source of material for rammed earth, quality and consistency need to be carefully checked. Materials not previously used should be tested for suitability, and sufficient resources (time and money) should be programmed into a project for material testing and selection. Care should be taken to ensure the stability of sub-soil constituents following excavation and use in rammed earth. Advice from a soils specialist, particularly on chemical and physical stability of constituents, may therefore be required. Depending on the nature of the project, testing might comprise laboratory work for grading and physical characteristics or simple qualitative testing by an experienced rammed earth contractor.

Factors influencing the selection of a suitable soil include:
- Available quality and quantity of in-situ soil
- Colour and texture of compacted material
- Transportation distances
- Storage of materials
- Opportunities for off-site fabrication

If the available in-situ soil is unsuitable, a blended or engineered material may be formed by granular stabilisation (combining different materials) in order to provide the desired mix characteristics. Aggregates with insufficient clay can be improved by the addition of powdered clay and silt mix for example. Alternatively, clay and silty soils can be improved by the addition of sand and gravel in suitable proportions. In this approach some use may be made of marginal in-situ materials, but construction is generally lengthened by the need to combine materials adequately in preparation. Costs of aggregates vary depending on source, use and transportation distance. Recent projects where blended soils have been successfully used include the AtEIC Building at CAT and the Chapel of Reconciliation in Berlin.

3.5.1 Soil survey

The bulk raw material of rammed earth construction is subject to natural variability and unfortunately not all sub-soils are suitable. Necessary resources, including time, must be allocated within a project towards soil selection. Observational survey of trends in local historical and existing earth buildings is a useful starting point to establish the likelihood of finding suitable material in the local area. UK soil survey and geological maps are a further useful resource for preliminary appraisal of in-situ material and for finding off-site sources of material. Material from local quarries and large public works projects are other possible sources.

Site investigation of materials for rammed earth should generally follow recognised procedures for civil engineering[18]. Materials may be sampled from bore-holes, trial pits, stockpiles and direct ground excavations. Sufficient samples must be taken to ensure that they are representative of the bulk material. Where large quantities of material are to be used it is important to undertake regular analysis to assess ongoing consistency.

Following sampling of materials, suitability for rammed earth construction is assessed on the basis of soil classification tests (grading, plasticity) and physical characteristics of prototype rammed earth specimens. Soil selection criteria outlined here are based on experience through published recommendations and will vary according to design preferences and requirements.

3.5.2 Soil classification tests

Though some soils testing will always be necessary with novel materials, the level of testing and analysis should reflect and be proportionate to the scale and complexity of the proposed works. Basic soil testing is recommended to determine grading (gravel, sand, silt and clay contents), plasticity, and organic matter content. Grading gives an indication of likely compaction and quantity of fines present. Plasticity indicates cohesive nature of the fines content. More detailed tests may be undertaken to determine level of soluble salts, soil mineralogical composition, including clay type(s), and pH. For earth building, grading and plasticity tests will often provide sufficient indication of clay reactivity and type (expansive, non-expansive).

Testing may follow procedures used for civil engineering classification[18], including sieve and sedimentation or pipette analysis, plastic limit, liquid limit, linear shrinkage, organic matter content and soluble salt content. Soil mineralogy may be determined by X-ray diffraction analysis.

Various simple field tests are frequently referred to in earth building publications as a means of assessing soil suitability[4]. These include: sensory tests for soil composition (visual, smell, touch); jar sedimentation test for volumetric soil composition; dry strength test for clay content; water retention test for indication of fines composition; thread and ribbon tests for clay content; and the shrinkage box test for plasticity. The reliability of these tests is variable and often dependent on the experience and interpretation of the user. Therefore, whilst field tests by experienced personnel may be useful for initial selection, where engineering design is required such analysis should not be treated as a substitute for a programme of laboratory testing.

3.5.3 Soil selection criteria for rammed earth

Wide varieties of sub-soils have been successfully used for rammed earth buildings. In general, soil for rammed earth should be well graded, containing gravel, sand, silt and clay fractions.

Ideally the soil should have reasonably high sand and gravel content, with some silt and sufficient clay to act as a binder and assist soil compaction. Maximum particle size depends on grading proportions and application. Increasing the proportion of larger particles increases the risk of surface defects such as 'boniness' and friable edges, and reduces compressive strength of the material. Maximum particle size is often limited to 10–20 mm, though particles over 50–100 mm have been used successfully. Suitable soils for rammed earth in general meet the criteria below or fall within with the upper and lower limit grading curves shown in Figure 36.

Sand and gravel content	45 to 80% (by mass)
Silt content	10 to 30% (by mass)
Clay content	5 to 20% (by mass)
Plasticity index	2 to 30 (liquid limit <45)
Linear shrinkage	Not more than 5%
Soluble salt content	Less than 2% (by mass)
Organic matter content	Less than 2% (by mass)

Sub-soils that only marginally satisfy the above guidelines, especially those demonstrating higher plasticity and shrinkage, should be treated with greater care.

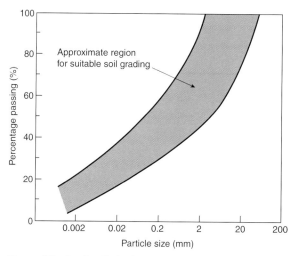

Figure 36 Grading limits for rammed earth soils

3.6 Physical characteristics

For the assessment of material suitability, physical testing of prototype specimens made from rammed earth and from stabilised rammed earth is recommended in conjunction with the soil classification tests previously described. Physical testing of rammed earth materials is also necessary to ensure that design requirements and specifications are met initially, and they may be checked later during construction. Design requirements for rammed earth materials vary depending on application.

3.6.1 Dry density

In assessment and preparation of materials, moisture–density relationships are determined using the heavy manual compaction test[17] (Appendix A.3.2). The test provides optimum moisture content and maximum dry density for material passing a 19 mm sieve. Larger aggregates present in construction may influence density and optimum moisture content. When more than 20–30% of the materials are greater than 19 mm, larger-size specimens should be used (Appendix A.3.1).

Specifications for rammed earth normally require a minimum of 98% of the heavy manual compaction test maximum dry density, though compactive effort of pneumatic compaction is higher than the modified test effort. In-situ testing of rammed earth to check the achieved density is difficult, so tests are typically undertaken on cylinders or cubes prepared by the contractor.

3.6.2 Compressive strength

Compressive strength of rammed earth is normally determined by testing cylindrical specimens in uni-axial compression (Appendix A.3.3). Dry unconfined compressive strength of rammed earth is normally in the range 0.5–4 N/mm^2. In contrast stabilised rammed earth can achieve strengths in excess of 10 N/mm^2 in less than 7 days (Appendix D). Cylinders are prepared in advance in the laboratory or on site. Specimens are often tested following drying under ambient laboratory environmental conditions. When more than 20–30% of the materials are bigger than 19 mm, larger-size specimens should be used (Appendix A.3.3).

Assumed design values should take account of ambient and likely worst-case moisture conditions under the design loading. For example, where significant loading is likely to be applied to newly rammed (damp) walls, lower values of strength and stiffness than the final dry values should be assumed. Compressive strength of moist rammed earth materials is likely to be at least 50% lower than the final ambient values.

Factors of safety are applied to design capacities based on material strengths to account for variations in materials and quality of work. Material properties are often determined from small-scale specimens, which do not include larger-scale constructional defects such as boniness (layers of under-compacted material) and out-of-plumb. A 'partial safety factor' for rammed earth of between 3.0 and 6.0 is used by this guide (Chapter 6 and Appendix C.1). The possible influence of boniness on the structural capacity of small cross-sectional areas such as columns should be considered.

3.6.3 Flexural and shear strength

The flexural tensile and shear strength of rammed earth is generally very low. Walls may require a minimum flexural tensile strength to resist lateral loads, though material self weight and other pre-compression loads will often be sufficient. Some shear strength may be assumed for racking resistance, though frictional resistance alone will often be sufficient. As well as the basic material characteristics, construction issues such as initial moisture content, extent of ramming and rate of drying, influence shrinkage and hence the bond between layers. Over-compaction is believed to be disruptive to the flexural and shear strength developed between compaction layers.

3.6.4 Deformation

Rammed earth walls deform owing to elastic displacement (shortening) under load, drying shrinkage, thermal expansion and possibly creep under sustained load. To minimise cracking such movements should be accommodated in design through appropriate detailing and the provision of movement joints. Where loadbearing rammed earth elements share load with other, often stiffer, elements deformation needs to be accommodated in design to avoid accidental overloading.

Elastic modulus
It is apparent from experimental data that there is no generalised relationship between compressive strength and elastic modulus of rammed earth. Reported elastic modulus values for rammed earth generally vary between 100 and 1000 N/mm^2. Therefore, where elastic modulus is important to design, for example where rammed earth sections share load with other stiffer structural elements, it is recommended that elastic modulus is determined from compression testing of representative samples of the materials (Appendix A.3.3).

Drying shrinkage
Following compaction rammed earth walls shrink vertically, laterally and longitudinally as the material dries from around 8–14% moisture content (by dry mass) at compaction to around 1–5% in ambient 'dry' conditions. The rate at which the material loses moisture and the final moisture content depend on many factors, including environmental conditions, shelter and material characteristics. As a general rule-of-thumb the rate of drying may be assumed to be approximately 25 mm wall depth per month to reach stable moisture conditions. Drying out can, however, be accelerated by the use of dehumidifying units for example.

The level of shrinkage depends on the clay content, soil grading, initial and final moisture content and rate of drying. In general, shrinkage will be less than 0.5%, though where its effect will be significant the drying shrinkage should be determined experimentally from specific material tests (Appendix A.3.4). In comparison with mud techniques, such as cob, the level of shrinkage is usually much lower. When walls dry unevenly, for example because of sheltering from the sun on one side, they may develop a lean towards the drier face. To prevent this, walls may require light propping until drying is complete (Figure 37).

Lateral drying shrinkage may be accommodated by the inclusion of movement joints. Vertical shrinkage is principally a concern where loadbearing rammed earth shares structural support with other elements, such as a timber or steel framework. In such cases quantifying the extent and rate of drying shrinkage is important. Fixings, such as wall ties, must also be able to accommodate the expected level of shrinkage, otherwise cracking may occur when restrained. Excessive

Physical characteristics

Figure 37 Propping of walls during drying

drying shrinkage and cracking may also undermine airtightness, acoustic separation and fire protection properties especially as walls are usually left unplastered. Higher plasticity, generally marginal quality, sub-soils require particular care in design and construction to minimise potential harmful effects of excessive shrinkage, such as cracking and loss of plumb, on drying.

Increases in moisture content during service life, flooding or leaking service pipes, may lead to significant swelling of the material. Localised softening of the material, for example due to a water leak, will be of most concern.

Thermal expansion
The coefficient of thermal expansion for rammed earth is in the range 4 to $6 \times 10^{-6}/°C$.

Movement joints
Movement joints in rammed earth walls may be provided to control deformation due to shrinkage and thermal fluctuations and to allow differential structural deformation to occur without damage to the wall. The horizontal spacing of vertical movement joints depends partly on material properties and partly on the design of the wall. Spacing of joints in rammed earth should tie in with the modular size of the shuttering system in use. Joint spacing also depends on the ground conditions and foundations provided. Further guidance on detailing of movement joints is given in Section 5.9.

3.6.5 Durability
Water-related weathering, including erosion by rain and freeze–thaw damage, and mechanical abrasion of surfaces are the primary agents of decay in rammed earth buildings. The

material's susceptibility to deterioration in the prolonged presence of water requires consideration throughout design, construction and maintenance. Moisture ingress into rammed earth walls may arise from rainfall, condensation, accidental flooding (burst water pipe), rising damp and from general building use. The collection of moisture at wall bases can be avoided by providing good drainage, damp proofing, surface protection, and a suitable eaves overhang (Section 5.4).

Rammed earth should generally be fully protected from weathering throughout its life. Rainfall causes damage through kinematic impact at the surface, washing out of fines and the cyclic swelling and shrinkage of the clay fraction. Further protection to external walls may be provided by protective coating or weather screens around the outside of buildings (Section 5.5). Exposure of rammed earth walls, with the expectation of erosion requiring ongoing repairs, requires knowledge of expected material performance. This knowledge is most likely to come from previous projects using the same material rather than laboratory testing. The rate of erosion of exposed earthen materials normally decreases with time as surface fines are removed and gravel content is exposed (Section 7.1).

Laboratory tests for erosion potential are difficult, at best approximate and at worst misleading, owing to the complexity and long-term nature of weathering that is to be replicated in a few hours or days. Water-spray erosion tests and abrasion tests, such as those described in Appendix A.3.7, can provide some relative indication of material performance, but cannot be confidently used to forecast actual performance. Good performance in accelerated tests may not remove the need to protect walls from weathering. In general, freeze–thaw deterioration of rammed earth walls will only be a concern when walls are damp, such as immediately after compaction. Protecting walls from damp will also provide frost protection.

Light and heavy mechanical abrasion of walls and surfaces during normal use needs to be considered. Rammed earth wall surfaces may dust when brushed and this may be inappropriate for the intended use. Protective coatings with vapour-permeable binders can reduce dusting and improve abrasive resistance. Dusting potential on drying generally increases with higher silt and clay contents. Rammed earth with higher shrinkage fines is more likely to dust than materials with less active clay constituents. Abrasion testing of materials (Appendix A.3.8), may be undertaken during the material selection process. Walls should be protected against abrasion damage, using temporary barriers for example, during the later stages of construction work.

3.6.6 Thermal properties

Thermal properties of rammed earth are strongly related to its density. Both thermal conductivity and capacity increase with material density. As a dense material, rammed earth has poor thermal insulating qualities. Its thermal conductivity is similar to that of building materials of comparable density, such as clay bricks, concrete and stone (Chapter 6).

The thermal mass, or capacity of rammed earth to store heat, is an important characteristic for its use in energy-efficient building design. Thermal heat capacity is the quantity of heat required to raise the temperature of one unit volume of material by one unit of temperature. Thermal heat capacity is the product of specific heat capacity, material density and element thickness. For example, a 300 mm thick rammed earth wall of dry density 1900 kg/m^3 has approximately the same thermal heat capacity as a 200 mm thick dense concrete block wall.

3.6.7 Fire resistance

Rammed earth is generally built using only inorganic non-flammable mineral-based materials. Occasionally natural fibres may be included, but their quantity is unlikely to impair performance in fire significantly. A typical solid rammed earth wall should generally achieve at least 90 minutes' resistance in fire, adequate for most likely applications. Rammed earth walls may be damaged from high-pressure fire hoses, so the consequences of accidental localised wall failure as a result of their use may be considered as part of the structural design.

3.6.8 Hygroscopic properties

Clay minerals present in the rammed earth are hygroscopic, absorbing and releasing moisture in response to changes in the surrounding environment. As previously discussed, this characteristic makes a significant contribution to regulation of internal environmental conditions, especially relative humidity. The application of protective coatings will in the main impair the hygroscopic behaviour of rammed earth walls. Potential swelling due to variation in atmospheric conditions is not considered significant for materials best suited to rammed earth.

4 Construction of rammed earth walls

This chapter is concerned with the practical application of on-site (in-situ) rammed earth construction. It covers everything from test panels and storage on site to formwork and protection of new works.

4.1 Preparation

4.1.1 Test panels

Building a test panel is a recommended basis for agreeing issues of surface quality and general detailing. Suggested test panel size is 500–1000 mm high × 500–1000 mm wide and 250–300 mm thick (see Figure 81). Where specific details such as joints are to be shown this requires discussion and agreement. It is recommended that the test panel is allowed to surface dry before inspection, as freshly rammed walls often look and feel quite different from work which has dried out. Some caution is, however, necessary with test panels, as large expanses of built wall may appear subtly different from a smaller area of test panel. Test panels also provide the opportunity to discuss and apply protective coatings and are recommended for novel or innovative loadbearing applications, to ensure confidence in the material mix and the method of application.

4.1.2 Site organisation

The extent of rammed earth works will determine much of the site organisation, as will other factors including speed of projected work and site situation. Depending on the size of the project, the quantity of both formwork and material will vary, and access to the site will affect both of these. Plant access is very important; in particular the soil delivery arrangements need to be scaled to the site and work carefully. Overly large machinery can be as disruptive to progress as too small.

4.1.3 Storage and preparation of material

Storage of material is all important; keeping stockpiled material dry is a constant theme. One simple way to reduce the problems of keeping material sufficiently dry in wet weather is to reduce the stockpile size wherever possible. Where material is being dug on site it may be preferable to leave the material in the ground until as late as possible before use, as long as the in-situ moisture is not above that required for compaction. Once material is dug up it can absorb rain much more readily than well drained sub-soil suitable for ramming. Dug material should be stockpiled on a well drained area and covered with tarpaulins or be inside suitable shelters. Suitable sub-soil will tend to dry quickly, so in good weather it is recommended that covers be removed to enable drying; adding moisture to material is much easier than drying it out in a hurry.

If soil needs screening it is much easier to do this at the point of purchase (a quarry) rather than on site. Where material is found on site and needs screening, there are a number of ways to do this, largely depending on the scale of the project. Small works can often be hand screened as the work progresses, while larger jobs may employ a number of different mechanical systems.

4.1.4 Formwork

Formwork in rammed earth construction is used as a temporary support during soil compaction. Like formwork for concrete it must have sufficient strength, stiffness and stability to resist the pressures it is subjected to during erection, placement of the material, and dismantling. However, unlike with concrete, rammed earth formwork can be removed almost immediately after compaction, enabling much faster re-use. In addition, formwork should be light, easy to assemble and disassemble, and robust enough to withstand repeated use and site practice. An efficient formwork system is therefore key to good productivity in rammed earth construction.

The choice of formwork to suit the job is very important, and many different choices exist. The first main choice is between moving formwork and static formwork. A moving formwork system requires setting up at each fill, with a team using one or two pieces of formwork throughout and building sections of wall much like laying very large bricks. This approach has been used extensively throughout the world and has led to a number of different basic formwork types. These can be summarised as either through-bolted (or clamped (Figure 38)) or cantilevered (Figure 39), while both have employed turnbuckles to achieve sufficient restraint in containing the pressures generated during compaction.

There are two major issues with moving formwork: the time taken for each set-up and the quality of the finished surface. The first is largely managed successfully by experience, and by having two boxes, one of which is filled while the other is

Preparation

Figure 38 Traditional timber formwork

Figure 39 Cantilevered formwork

stripped and re-erected. Surface finish is not usually a problem where a plaster or other protective coating will be applied to the work.

Static formwork, using reinforced concrete formwork systems, is often used where a high-quality finish is required. Static formwork requires setting up at ground level, but subsequent lifts are clipped onto the first layer, which increases speed and quality of the surface finish (Figure 40). Modular panels are clipped together and strengthened by clamps, ties and supports. Full-height formwork systems, including extended rammers, have been developed to minimise time spent erecting formwork in order to increase productivity.

There are a number of specific systems designed and built, particularly in Australia and the United States of America, for rammed earth. However, commercially available concrete systems are fast becoming much more user friendly for rammed earth (Figure 41). In the past concrete formwork has not been sized well for rammed earth, which requires a high level of manoeuvrability. As each set-up is filled another lift of formwork has to be clipped into place and levelled. Handling requires formwork to be both light enough to lift yet robust enough to contain the forces generated by ramming.

As well as proprietary systems, bespoke timber-based systems have also been used successfully (Figure 42). Timber (plywood) sheathing is combined with either timber or metal strong backs (walers and soldiers). Timber provides particular

Figure 40 Australian proprietary static formwork

Figure 41 Proprietary concrete static formwork system

Figure 42 Timber formwork

Figure 43 Timber formwork for curved wall

Figure 44 Through-bolted formwork

flexibility for curved forms (Figure 43). Timber formwork is suited to both moving and static systems. Though offering much greater flexibility than proprietary systems, productivity with static timber formwork is generally lower. When rammed earth workers are inexperienced there is also a greater risk of formwork system failure through excessive deformation, or possibly even collapse, under pressure.

Formwork stiffness, to limit deflections during compaction, is often ensured by providing through-bolts (Figure 44). Bolts should be spaced sufficiently close together (500–1200 mm) to limit formwork deflection without hindering compaction. On removal of the formwork, bolt-holes are normally patched with matching earthen material. However, where a clean wall surface, without bolt-holes, is desired, formwork deformation is limited by increasing the stiffness of the formwork system, using external ties and clamps, and using external props.

4.1.5 Openings

Openings may be formed by using temporary block-outs or incorporating supporting lintels during construction, though arguably the simplest way to form openings is to extend the opening to the full height of the wall. Openings cannot easily be cut into a solid wall after construction because of the high density and strength of rammed earth. In blocking-out, robust boxes of plywood are made to the required dimensions of the openings and inserted in the formwork at the location where the openings are to be formed. After ramming, the block-outs are removed to provide the opening.

4.2 Building

4.2.1 Mixing and placing

A significant difference between rammed earth and most other wall building materials is that rammed earth moves straight from a loose prepared material to a post-compacted in-situ wall with very few intermediary stages. Materials are prepared before construction: sub-soils are excavated and screened, large aggregated lumps of material are broken down or screened out, materials are combined, and the constituents are thoroughly mixed together. Achieving a uniform and consistent mixture is very important. In situations where water has been added, more uniform moisture distribution throughout the soil is usually achieved if the material is left to stand, without drying, overnight; a process sometimes referred to as maturation.

Once the constituents have been prepared, sufficient water is added and mixed in to bring the mixture to its optimum moisture content. The material should be at its optimum moisture content when placed into the formwork in preparation for compaction. Stockpiled materials should therefore be protected to prevent excess wetting, as it is much easier to add water than to dry out materials that are too wet. Soil moisture content during construction is checked at regular intervals, qualitatively, by experienced rammed earth practitioners using the drop test (Appendix A.3.2.3). Alternatively, rapid quantitative measurements of soil moisture content may be made on site using a carbide moisture meter[19] or by weighing small, but representative, samples before and after drying in a microwave oven.

Various methods are used in mixing materials. Very small amounts, such as those used in repair, may be prepared by hand, though this is not a practical approach for larger-scale works. Mechanical mixing is the most common approach using either concrete, screed or mortar mixers or using construction or

Figure 45 Small forced-action screed mixer

Figure 46 Pan-style concrete mixer

agricultural plant. The chosen method of soil mixing should not restrict the rate of wall construction.

Mechanical mixers should be able to break down small aggregated lumps of soil and provide a uniform consistent mixture of solids and water without balling. No single mixer type is preferred as it depends on the scale of works and characteristics of the materials. Suitable mechanical mixers include forced-action (screed) mixers and pan concrete mixers (Figures 45 and 46). Rotating drum mixers have the greatest tendency to ball materials.

Skid steer front end loaders, such as the 'Bobcat', are commonly used by rammed earth contractors for mixing and lifting materials into the formwork (Figure 47). Materials are mixed by lifting and turning on the ground using the loader. This method is probably the fastest and is particularly well suited to low-cohesive soils such as those commonly used in stabilised rammed earth (Appendix D). The ground on which mixing is undertaken should be free from contaminants. Agricultural and garden cultivators provide another method for mixing materials (Figure 48). They may be hand-operated or mounted on a small tractor.

The mix may be placed into the formwork in a number of ways. On larger-scale work a number of different technical solutions exist for placing material. Front-end loaders of various sizes have been used to good effect, but getting the size right for

Building

Figure 47 Skid steer loader

Figure 48 Rotavator mixer

the job is important. Cranes have occasionally also been used with concrete placing systems; these are used in conjunction with a digger or front-end loader for higher access.

The placing method affects the appearance of the material when the formwork has been removed, and an element of manual work just before ramming to ensure that the earth is well distributed in even layers, increases the quality of finish of the final wall. However, where walls are to receive a plaster or render this may be less desirable. Placing material to produce a smoother finish is achieved through disaggregation, placing the material against the formwork face to encourage larger-aggregate particles to move to the centre of the wall.

The depth and level of the fill should be checked before ramming starts; even depth of fill ensures even compaction and a regular appearance on the face of the wall.

4.2.2 Rammers

Ramming may be carried out manually or mechanically in a number of ways. Typically manual ramming (Figure 34) will be carried out on a shallower depth of fill (less than 100 mm), though the shape and weight of the rammer does affect the degree of compaction. To optimise applied compactive effort rammer heads are generally kept small. A typical pneumatic rammer head is circular with a diameter of 150 mm (Figures 33 and 49). Traditional manual rammers often have a pointed bottom face, rather than flat as with pneumatic heads, to improve compaction. Other means of ramming have also been used, such as a small sheeps-foot roller, though these need clear access into the formwork (Figure 50).

Figure 49 Pneumatic compaction of a stabilised rammed earth wall

Figure 50 Compaction using sheeps-foot roller

4.2.3 Ramming

Soil at the correct moisture content will readily compact under a rammer. Soil that is too wet will not compact into a fixed matrix, but will continue to move around under the rammer. Material that is too dry may be harder to assess in the formwork, but generally material that throws up dust when compacted is too dry.

Suitable sub-soil at the right moisture content will compact readily and should be compacted first against the faces of the formwork. Once the material has been compacted, and will not compact further, it is counter productive to continue ramming and this may in fact lead to horizontal delamination. Rammed earth contractors often refer to a change in sound or 'ringing' of hand or pneumatic compactor heads at this stage.

4.2.4 Flow of work

Rammed earth is a relatively quick way to build if the flow rate is right. This means always having an empty box set ready to be filled. In this way the mixing and placing flow are constant as is the rate of ramming. When formwork is not available to be filled, work stops and the flow rate of the whole team suffers.

4.2.5 Formwork removal

Removing formwork is a key moment in the construction of a wall. Hasty or ill-considered removal can easily damage the soft fresh material. Formwork should always be slid along the surface of the newly built wall to break any seal between the two surfaces. Once the formwork moves easily it is safe to pull it away. This is true of both moving and static formwork, particularly where the face of the form is smooth and sealed. Open-grained timber formwork tends to stick less, as do more textured surfaces.

With static formwork systems it is important to remove the formwork sequentially from the top, removing the complete top layer before working down the wall. This is true even where static formwork has been slip formed up the wall leaving new work exposed below during the course of the work.

4.2.6 Construction sequence

Whatever formwork system is being used, the sequencing of the work is important to the speed of production as well as the end result. Moving formwork typically moves horizontally around the site, building first one complete course, then a second and so on. Sequencing the moves to allow access for the material around the site is important. The other sequencing issue, which moving and static formwork share, is shrinkage of the compacted earth. Where this may be a particular concern, allowing a few days or longer between vertical joints may be advisable. This is more prominent in static formwork where the vertical joints are longer and leads to a 'hit and miss' approach. With 'hit and miss' every other wall panel is left unbuilt initially. After a number of built panels have had a chance to lose moisture and shrink, the unbuilt sections can be 'filled in', reducing the tendency of the panels to shrink away from each other. As with moving formwork, planning the sequence to allow access to the work throughout is important.

4.2.7 Movement joints

Movement joints vary depending on whether moving or static formwork is being used. With moving formwork the amount of shrinkage between the vertical ends of panels tends to be reduced by shorter fill lengths and then by interlocking of the subsequent lifts built above. Because of this and the more random nature of the placing of formwork, heavily detailed joints on the surface are generally avoided. Where more than simple butt joints are required a key piece needs to be added to the end stops to allow formed walls to interlock (Figure 51). Detailing of movement joints is discussed in Section 5.9.

Where necessary, to provide an acoustic barrier and fire resistance, it is advisable to include a foam or similar strip with intumescent properties between vertical pieces to reduce any loss of airtightness due to shrinkage. This may be used in conjunction with the keyed end and should not be visible from the outside (Figure 51(c)).

Horizontal joints are not detailed but are simple compression joints, effective because the higher lift is rammed onto the lower and is then held by its own weight. Day joints are not seen as problematic structurally but removing dry edges before recommencing ramming is advisable where a smooth finish is required. The upper surfaces of ongoing works may be protected at the end of a day's work to prevent excessive drying or wetting.

Using static formwork results in fewer vertical joints between built sections but produces long vertical junctions. Earth walls can be formed into 90° corners, but these are susceptible to mechanical and water damage. To reduce the risk of this walls are generally formed with chamfered corners. Once larger chamfers have been formed a smaller jointed detail then allows wall sections to interlock neatly (Figure 51).

As with the moving formwork, vertical joints may open slightly owing to shrinkage. To mitigate the effects of this an expanding foam or similar jointing piece may be fixed against an existing wall end and be built into the joint.

Building

(a)

(b)

(c)

Figure 51 Movement joints

Building up to existing rammed earth is a long established practice. Most types of other building material have been used next to rammed earth, including block and brick, cast concrete, stone masonry, timber and steel. Generally no special problems have been found with this, but if fixings need to be made to attach timber or steel for instance, then standard fixings need to penetrate deeper than with fired brick or blockwork masonry (Section 5.7). Building earth up to existing structures requires particular consideration. The pressure generated from a 2 m to 3 m high wall may be considerable and temporary propping may be necessary to ensure that the existing structure is not permanently damaged or displaced.

On the whole it is preferable not to tie new rammed earth works to existing structures, as ties restrict movement, the bond of ties to unstabilised earth is unpredictable and the durability of embedded items is unclear. However, when considered necessary, ties to existing structures should be able to accommodate forces developed from restrained horizontal shrinkage, whilst allowing vertical shrinkage to occur relatively unrestrained, for example wall ties attached to a sliding rail support.

4.2.8 Protection to new works

Once the formwork is removed the new wall starts to dry out. The weather, the initial moisture content, wall design and thickness all play a part in how long the drying process will take. This is the most crucial time for the wall, it is at its most vulnerable in the first weeks after construction, or months in inclement or colder weather. So there is a conflict between leaving the walls open to dry and covering them up to protect them from heavy rain.

While work is under way plastic covers may be used to provide either a simple capping or full coverage of the walls (Figure 52). However, great care must be taken with covers because plastic may concentrate flows of water onto a localised area of wall, and these concentrations of water may wash finer particles out faster than if no cover were provided. As walls dry they generally become more resistant to rain.

The sooner a roof or permanent cover can be put in place the better. If a very heavy roof is planned, walls may need more time to dry before the full load is imposed. The construction of roofs should begin as soon as possible. In very large structures or where wall building is slow for programming reasons, roofing

Figure 52 Protection of new works

should begin before the walls are complete. This sequential method of building should itself be programmed to prevent large areas of wall being built a long time before the roof will be put on. This is one of the few ways in which rammed earth differs markedly from other materials.

In more complex contractual situations clear contractual responsibility needs to be established. While earth walling is still being built it is generally the earth builder's responsibility to keep walls protected. Once wall building is complete responsibility must be established between the main contractor and other subcontractors, including the roofer, to protect and cover the walls.

4.2.9 Surface finish

Since rammed earth is formed by ramming earth against a fixed formwork, it follows that the face of the formwork plays an important role in the quality of the finish of the wall. Smooth formwork produces a smoother finish, coarser timber formwork can give an interesting textured surface and so on. Care should be taken though with very smooth formwork not to pull the formwork away from the face as this may lead to material being plucked off. Formwork should always be slid along the surface of the wall before removal to break the seal between the two faces.

The initial appearance of the wall will change as the material dries. This may happen very quickly or over days or even weeks. The appearance of the raw material of the newly rammed wall may not be a good indication of the finished material. The colour will lighten with drying, though it can be darkened again with the use of some surface finishes.

5 Details for rammed earth construction

This chapter sets out recommended construction details for footings and wall bases, openings and supporting beams, roof protection, provision of services, fixings to rammed earth walls, and protective coatings, and gives guidance on thermal and acoustic insulation.

5.1 General

Rammed earth is a solid form of wall construction. Detailing of rammed earth structures is governed by the material's susceptibility to deterioration in water combined with its relatively low strength and abrasion resistance, and relatively high thermal conductivity. General design issues to improve durability of rammed earth include:

- Ensuring that adjoining structures are designed and built to shed water away from the rammed earth wall
- Use of protective coatings or weather screens in areas subject to high winds, such as gable ends, where simple protection from roof projection may prove inadequate
- Generally avoiding building walls in sites prone to flooding or standing water, though risk may be minimised by extended plinths. Walls should always be built on raised footings
- Allowing excess moisture to evaporate from walls to minimise the build up of damp within the fabric
- Designing to reduce the likelihood of vandalism and incidence of deliberate abrasion damage to wall faces

5.2 Footings and base details

Footings for rammed earth walls follow design and construction provisions for similar solid masonry walls. Protection from water damage, moisture ingress and, in appropriate areas, radon gas, are the governing criteria for wall footing details. Given the comparatively low stresses at the base of rammed earth walls, simple mass footings, such as concrete, limecrete, rubble masonry or even cement-stabilised rammed earth, are likely to be suitable for many situations. Construction of a plinth upstand (stem wall) as part of the footing detail, greatly assists subsequent construction of the walls by defining the plan position and line for setting the formwork.

As with all solid walling there is potential for water to rise up from the ground and through the base of the wall. Though in traditional earthen construction damp-proofing barriers were not provided, modern building practice requires the provision of a continuous damp-proofing barrier in accordance with current standards and codes of practice. Where possible the damp-proof course should be made continuous with the damp-proof membrane, as is common practice. Radon barriers should also be provided at wall bases and beneath floors in new buildings in affected areas.

The damp-proofing barrier is often provided along the interface between the footing or plinth and the base of the rammed earth wall (Figures 53 and 54). In such cases the damp-proofing material should be capable of withstanding ramming without damage. Heavy-duty plastics-based damp-proof coursing materials are commonly used. However, to limit the possible risk of moisture build up and damage at the interface between the rammed earth base and impermeable damp-proofing barrier, the barrier might be placed lower down the plinth beneath a course of bricks or similar (Figure 54).

Figure 53 Damp-proof course

Footings and base details

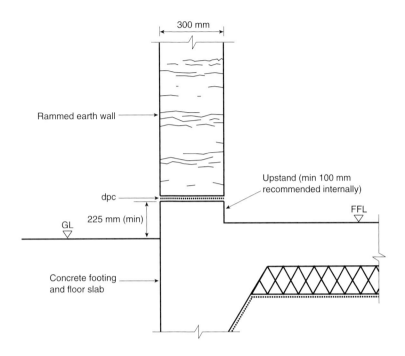

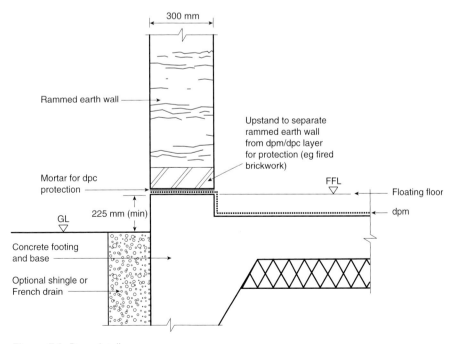

Figure 54 Base details

Figure 55 Water damage at the base of a wall

Rammed earth is vulnerable to water splashing on the external face at the base, which may cause pitting and erosion over time (Figure 55). The current Building Regulations are concerned that splashed water against the base of the wall causes penetrating damp and therefore have a general requirement to place the damp-proof course above the level of the external ground surface, to ensure water proofing in this splashing zone. It is recommended that rammed earth walls are built on upstands or plinths at least 225 mm above the exterior ground level.

Drainage is recommended at the base of external walls to limit moisture build up or ingress. Rainfall may be deflected away from the wall by grading the ground away from the building. Shingle drains or soft surfaces that minimise the risk of splash-back from falling rain are also recommended beneath the eaves (Figure 54). Vegetation should be planted away from walls to prevent the build up of moisture or water damage from irrigation.

5.3 Openings and supports

5.3.1 Provision for openings

Openings in rammed earth may be formed by creating full-height or partial-height sections when building individual freestanding panels of solid earth, by using boxed shuttered forms or by using structural lintels. Detailing window and door openings up to full height of the wall avoids the need for structural support within the rammed earth (Figure 56). Arched and flat openings formed by box shuttering inserted inside the wall formwork are an effective means of providing openings over modest spans up to 1.0 m (Figure 57). Lintels may be formed from solid timber, concrete, stone or other suitable materials. Lintels will normally be full width and thus will be visible on both faces of the wall. Bearing length for lintels should normally be at least 300 mm for spans up to 2 m to avoid over-stressing weak edges. Lintel design should otherwise follow general practice used for openings in weak masonry.

When a lintel is used to provide horizontal support over an opening, it is normally placed in position within the formwork during soil compaction to allow for possible settlement both during and after ramming. Where drying shrinkage in the material is expected to be significant, an allowance for this

Figure 56 Full-height opening between panels

Figure 57 Arched opening

Details for rammed earth construction

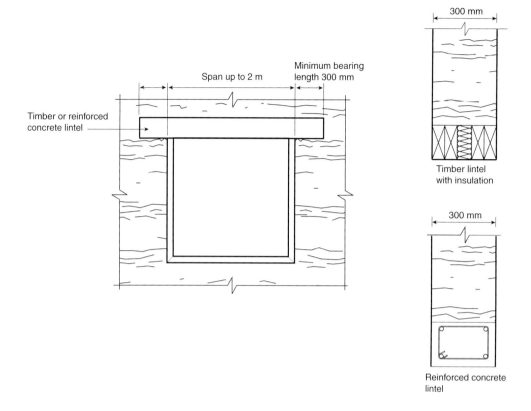

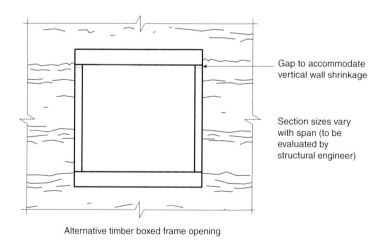

Figure 58 Opening details

Openings and supports

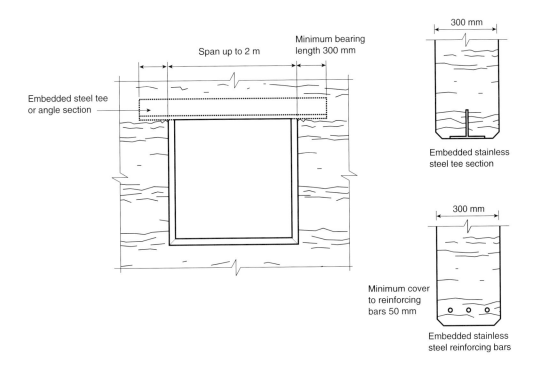

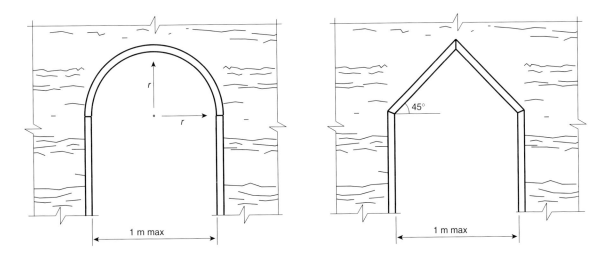

Figure 58 (continued) Opening details

movement should be made for subsequent incorporation of the window frames beneath the lintel (Figure 58). Expected material drying shrinkage may be determined in advance in accordance with the test procedure in Appendix A.3.4.

5.3.2 Wall plates

It is common practice in rammed earth construction to provide a wall plate or ring beam continuously along the top of the walls. Wall plates enhance stability when walls are subject to high lateral loads and provide an interface between the wall and roof for connection, anchorage and distribution of the roof loadings. Wall plates are usually either timber or reinforced concrete. Reinforced concrete bond beams are more usually provided where high horizontal forces are expected. As with concrete lintels they can be either pre-cast or cast *in situ*. Timber wall plates are fixed to the rammed earth wall using anchorage bolts or ties (Figure 59). The bond between in-situ concrete and rammed earth may be enhanced using ties, embedded anchors or 100–150 mm galvanised nails embedded into the top of the wall.

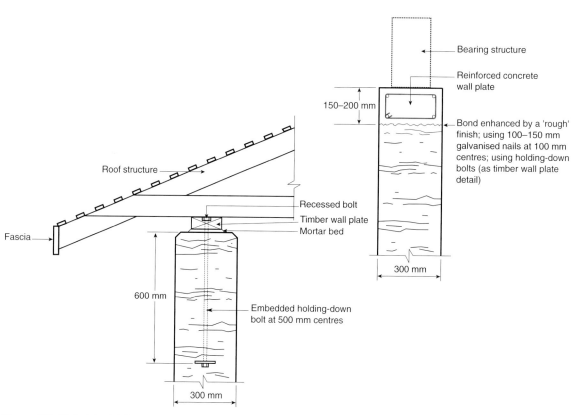

Figure 59 Wall plate details

5.4 Protection given by roofs

Eaves details should take into account the wall below the eaves and the protection required because of its exposure. Eaves extensions provide an important means of protecting external wall surfaces from rain (Figure 60). Detailing of the plinth at the base of the wall should be consistent with that for the roof, providing maximum possible protection. Eaves for a 3 m high wall should normally extend a minimum 400 mm beyond the vertical line of the wall at its base, though for the most severely exposed sites greater protection of up to one-third of the wall height may be provided for what is generally considered to be full protection from rainfall[20]. Eaves protection requirements will vary according to site exposure and available shelter for the wall.

Where the wall is to be over-clad, in timber for example, general rules for timber weathering apply. Where there is a fascia and soffit there is a risk of water blowing against the wall face; this should be avoided by robust detail, or protective coatings might be applied to the face of the wall in these areas.

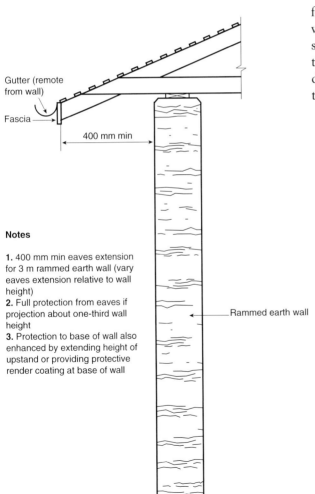

Notes

1. 400 mm min eaves extension for 3 m rammed earth wall (vary eaves extension relative to wall height)
2. Full protection from eaves if projection about one-third wall height
3. Protection to base of wall also enhanced by extending height of upstand or providing protective render coating at base of wall

Figure 60 Eaves details

5.5 Protective coatings

Rammed earth walls are on the whole best left without any applied protective coatings. However, protective coatings can be necessary or desirable on general internal surfaces, to limit dusting, on internal surfaces in wet areas and on exposed external surfaces. The necessity and success of protective coatings vary according to material qualities and design details. The effectiveness of these protective coatings depends on many factors, including material characteristics of the rammed earth and method and timing of application. Whilst there is considerable experience and success with some methods of protection, such as lime renders, there is generally much less experience and knowledge of transparent coatings. Protective coatings should never be used as a universal panacea for durability in the absence of other measures. Once applied some protective coatings will require regular re-application, possibly every 1 to 5 years as part of routine maintenance, depending on materials and exposure.

It is important that protective coatings do not form a seal which prevents transfer of water vapour. Though walls should remain water vapour permeable, protective coatings will generally impair moisture transfer including the hygroscopic regulation of relative humidity. Any water that ingresses into a wall, through rising damp or damaged pipes, should be able to disperse throughout the wall and escape through evaporation. Walls should also be well ventilated to prevent build up of damp areas further. Protective coatings should also be sufficiently flexible to tolerate movements in the wall, including further drying shrinkage. Brittle coatings are best avoided. Cement-rich renders should never be used on rammed earth walls.

Renders and plasters hide the natural rammed earth wall finish. Maintaining the original aesthetic has therefore been the focus for using a variety of transparent protective coatings. Protective coating materials used, with varying degrees of success, include liquid silicate solutions, proprietary water repellents and sealants (including polyvinyl acetate (PVA) and silicone-based emulsions), a variety of water-based sealants normally used on concrete surfaces, and natural oils, such as linseed oil. Particular regard should be given to preserving vapour permeability of the surface, albeit reduced in comparison with the untreated material. As experience with many protective coatings is limited and rammed earth materials varied, trial test panels are generally recommended prior to main application to check colour, texture, bonding and performance in advance.

Protective coatings

Recent experience with some protective coating methods such as sodium silicate spray, has highlighted problems. Effectiveness of protective coatings depends on the depth of penetration into the surface. Sodium silicate treated walls are often characterised by the development of a thin brittle surface skin, around 2–4 mm thick, representing the penetration depth of the protective coating (Figure 61). As long as the protective coating remains integral it provides protection against surface erosion. However, failure of the protective coating can often lead to preferential and localised higher rates of weathering (Figure 62).

For external walls lime renders provide a suitable vapour-permeable weather-resistant protective coating. The render may be applied directly to the wall face or onto lath mesh supported on battens as part of external thermal insulation measures (see Figure 67(a)). Though render coats are generally around 12–25 mm thick, they may incorporate lightweight aggregates in layers of up to 100 mm to enhance thermal insulation (Figure 23). Lime and clay plasters may also be applied to good effect on internal surfaces (Figure 63). Limewashes, applied directly to the wall or onto renders, can provide additional wearing protection on both internal and external surfaces. Render and plaster coats may be applied

Figure 61 Peeling failure of sodium silicate protective coating

Figure 62 Preferential weathering of sodium silicate treated wall, exacerbated by under compaction

(John Renwick)

Figure 63 Clay plaster, Woodley Park Sports Centre

manually or, increasingly commonly, pneumatically. Advice on lime and clay renders and plasters is available elsewhere[4,6].

Drying shrinkage should ideally be completed before lime renders are applied, to prevent the possibility of subsequent failure. However, the timing of modern construction projects does not always allow this, so walls should be left to dry as long as possible. Drying can be accelerated by the use of de-humidifying units. If further shrinkage of the wall is expected the applied protective coating should be able to tolerate any further shrinkage of the rammed earth, not lose bond with the earth and not significantly impair surface vapour permeability. If the

Protective coatings

Figure 64 Movement joints in lime render

render or plaster is directly applied to the wall surface, horizontal movement joints may be provided to accommodate further shrinkage (Figure 64).

To preserve vapour permeability ceramic tiling should generally be applied to a raised rigid surface supported on battens and studs off the rammed earth wall.

5.6 Services

The thick monolithic nature of rammed earth walls provides sufficient space for incorporation of electrical services within the wall (Figure 65). This, however, should be discussed and planned with the rammed earth contractor, as the inclusion of vertical conduits may hamper the ramming process; where this is the case it should be avoided. Back boxes can be incorporated into the shuttering (Figure 66). Alternatively service runs can be located on the surface of the wall within a conduit or hidden within wall cladding. Rammed earth can be readily chased with an electrically powered grinder or similar. However, plastering and patching over chasing in rammed earth is generally unsatisfactory because of the poor-quality finish often achieved.

Water pipes are generally not permitted to be placed within walls and in general should be placed away from exposed rammed earth surfaces. However, underfloor heating pipes have been successfully incorporated into rammed earth floors.

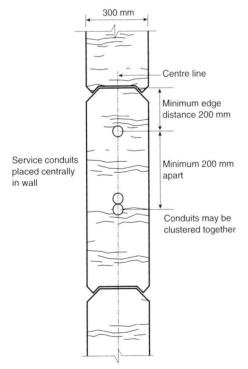

Figure 65 Plan view of embedded electrical services

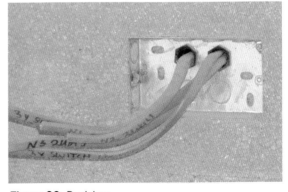

Figure 66 Back box

5.7 Fixings

Non-structural fixings are similar to those used for weak masonry of similar strength (see Helifix website *http://www.helifix.co.uk*). In recent years some experience has been gained using stainless steel tie attachments to traditional cob walls. For lightweight fixtures (photo-frames, paintings) masonry nails, screws and hooks with plugs can be used. For heavier fixtures, proprietary mechanical anchors, such as Rawlbolts® or Helifix® stainless ties (8 or 10 mm diameter), or some epoxy resin anchors are recommended. Ties should be inserted at least 150–200 mm into the wall and normally inset at least 150 mm from the edge of the wall. Strength can be improved by inclining the fixing in the wall. Pullout loads up to 1 kN may be expected for 8–10 mm mechanical anchors, though boniness and edge distance will also affect this performance. Where anchoring directly to the rammed earth is not suitable, timber battens embedded in the wall may be used for the heaviest attachments. Further advice should be sought from specialist fixing suppliers.

5.8 Thermal insulation

External and internal thermal insulation options are shown in Figure 67. Insulation materials in use with rammed earth should in general be vapour permeable. However, where impermeable materials are used, such as rigid foam-backed plasterboard for internal insulation, a 25–50 mm ventilated cavity should be provided to prevent the build up of condensation on cold surfaces. Normal robust details for solid wall construction apply to ensure the elimination of cold spots and bridges. Movement joints should be detailed with expanding fill material to ensure airtightness (Figure 68).

5.9 Acoustic separation

Rammed earth as a dense material provides an excellent barrier to noise and will attenuate a wide variety of frequencies. However, Approved Document C[13] is only concerned with an overall described attenuation. Rammed earth walls should be detailed as similar solid masonry wall construction, with appropriate discontinuous arrangement of adjacent structure, services and envelope at boundary walls separating properties. Movement joints may break the acoustic barrier and so should be positioned carefully and filled with an expanding fill material (Figure 68).

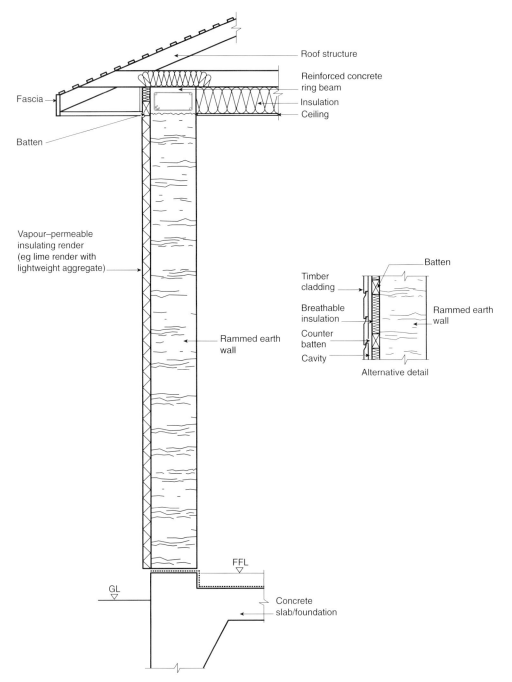

Figure 67(a) External insulation details

Acoustic separation

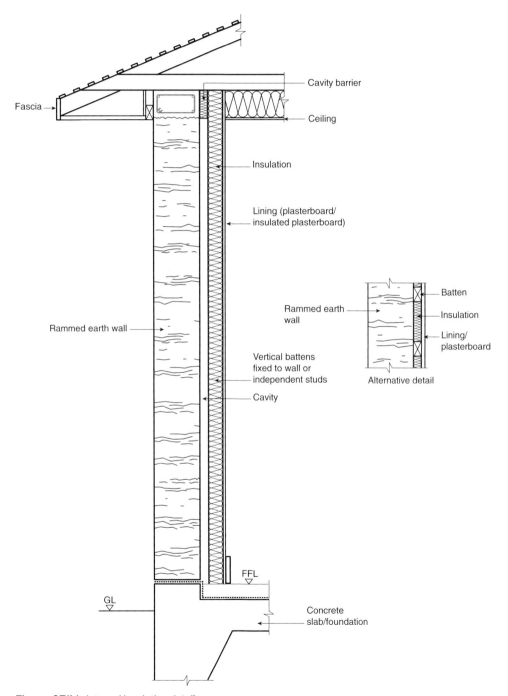

Figure 67(b) Internal insulation details

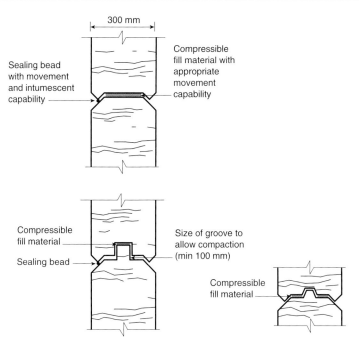

Figure 68 Typical vertical movement joint details

5.10 Construction tolerances

Recommended and reasonable construction tolerances for newly built rammed earth construction are outlined in Table 1. These are comparable with those required of masonry construction.

Table 1 Construction tolerances for rammed earth construction	
Horizontal position of any rammed earth element specified or shown in plan at its base or at each storey level	±10 mm
Deviation within a storey from a vertical line through the base of the member	±10 mm per 3 m of height
Deviation from vertical in total height of building (from base)	±15 mm per 7 m of height
Deviation (bow) from line in plan in any length up to 10 m	±10 mm per 5 m of length
Deviation from vertical at surface against which joinery is to be fitted	±10 mm
Deviation from design wall thickness	±10 mm
Position of individual rammed earth formwork panels	± 5 mm

6 Engineering design of rammed earth walls

This chapter sets out guidelines for the engineering design of loadbearing and non-loadbearing rammed earth walls. Design properties are provided together with rules for simple design. Further guidance for structural design is given in Appendix C. Retaining walls are not considered.

6.1 Design requirements

Rammed earth walls should aim to remain serviceable throughout their expected design life and not to suffer undue deterioration from weathering, accidental damage, animal infestation and general 'wear and tear'. As well as adopting preventative measures, such as protective coatings and design details to limit weathering, this design requirement may be met through a programme of ongoing maintenance and repair to walls.

Rammed earth walls should have sufficient strength and stability to withstand the actions (loads) that can be reasonably expected during their design life. Properties of rammed earth change with moisture content levels, so influence of raised moisture levels on strength and stability should be considered in design.

Rammed earth walls should meet all other design requirements, including quality of initial and 'final' surface finish, acoustic separation and thermal insulation.

6.2 Properties of rammed earth for design

6.2.1 Structural properties
Dry density
In the absence of specific material data, the characteristic dry density of rammed earth for assessment of self-weight design loadings should be taken as either 1750 kg/m^3 (beneficial, such as resistance to overturning) or 2250 kg/m^3 (unfavourable dead loading).

Compressive strength
The minimum characteristic unconfined compressive strength for all rammed earth construction is recommended to be 1 N/mm² when tested in accordance with the procedure outlined in Appendix A.3.3. Materials should be tested at ambient environmental conditions similar to those expected in service conditions.

Flexural tensile strength
Flexural tensile strength of rammed earth should not be relied upon in design without testing (Appendix A.3.5).

Shear strength
A coefficient of friction for rammed earth may be taken as between 0.20 and 0.30. The basic shear strength of rammed earth should not be relied upon in design without testing (Appendix A.3.6).

Elastic modulus
In the absence of test data an elastic modulus in the range 100–500 N/mm² may be assumed. Elastic modulus may be determined from measuring axial deformations during compressive strength testing (Appendix A.3.3).

Drying shrinkage
Where required in design, the drying shrinkage of rammed earth is to be determined in accordance with the procedure set out in Appendix A.3.4.

6.2.2 Thermal properties

Thermal conductivity
Thermal transmittance (U) and thermal resistance (R) values for rammed earth walls may be estimated on the basis of the thermal conductivity values given in Table 2.

Table 2 Thermal conductivity of rammed earth

Dry density (kg/m³)	Thermal conductivity (W/mK)
1400	0.60
1600	0.80
1800	1.00
1900	1.30
2000	1.60

Thermal heat capacity
A 300 mm thick rammed earth wall of dry density 1900 kg/m³ has a thermal heat capacity of 450–570 kJ/m²K.

6.2.3 Acoustic properties

The acoustic insulation quality of solid wall construction is closely related to its density or 'surface mass' (m', kg/m^2). The airborne sound insulation value, R_w, for internal rammed earth walls is given[21] by:

$$R_w = 21.65 \log_{10} m' - 2.3 \qquad \text{(where } m' \geq 50 \text{ kg/m}^2\text{)}$$

For 300 mm thick walls R_w will therefore generally vary, as a result of material density changes, between 54 and 59 dB, comfortably meeting the current minimum requirement of Part E of the Building Regulations[12] for internal walls of $R_w = 40$ dB.

Part E of the current Building Regulations requires the airborne sound insulation of separating walls to be tested *in situ*, as it is dependent on other construction materials and details.

6.2.4 Fire resistance

A solid rammed earth wall should generally achieve at least 90 minutes' resistance in fire.

6.3 Simplified design for structural adequacy

For most one- or two-storey rammed earth buildings, a simple structural design approach based on limiting geometry and axial compressive stress will be sufficient. For situations where such an approach is not appropriate a more rigorous methodology for checking wall capacity in vertical compression, with and without bending, in vertical flexure and in shear, based on the principles of structural masonry design, is outlined in Appendix C. Where rational engineering principles can demonstrate that thinner or more slender walls will be adequate, the requirements set out below need no longer govern. In the selection of minimum wall thickness, constructional and other design issues, such as acoustic and thermal performance, are also likely to have a significant influence.

Design geometry

Design thickness of a rammed earth wall should normally be taken as the overall thickness less any depth for recesses, holes or chasing. Design properties, such as cross-sectional area, section modulus and moment of inertia, should be based on the design thickness.

Unsupported clear height of a wall should be taken as the clear distance between lateral restraints.

Minimum wall thickness

The minimum thickness of rammed earth walls is primarily governed by ease of construction. For this reason rammed earth walls are normally at least 300 mm thick.

Maximum wall slenderness

The maximum recommended unsupported clear height between effective lateral supports for both non-loadbearing and loadbearing rammed earth walls is 8 times the minimum thickness for free-standing walls and 12 times the minimum thickness for walls restrained laterally top and bottom (Figure 69).

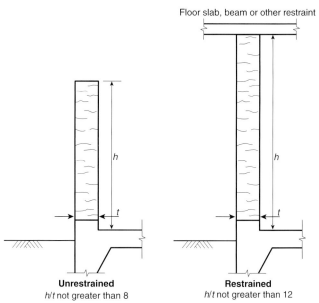

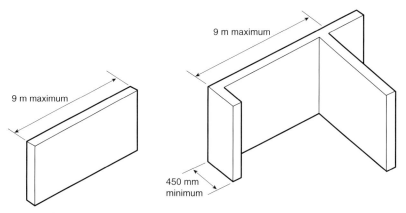

Figure 69 Limiting thickness for free-standing and supporting walls

Simplified design for structural adequacy

Compressive strength

For loadbearing rammed earth construction the minimum characteristic unconfined compressive material strength, tested at ambient in-service conditions in accordance with the procedure outlined in Appendix A.3.3, is recommended to be 2 N/mm².

Alternatively, the design compressive stress in a rammed earth wall, under the least favourable ultimate limit state loading conditions, should not exceed a value given by the measured or assumed unconfined material compressive strength divided by a suitable partial safety factor of at least 3.0–6.0. Advice on selecting a value for the partial safety factor is given in Appendix C.1.

Provision of openings

The provision of openings in rammed earth walls should not impair the structural robustness of the wall. General recommendations for openings are as follows (Figure 70):
- The total combined horizontal length of openings in a wall should not exceed one-third of the total wall length
- The distance between openings for a loadbearing wall of minimum thickness should be at least 600 mm
- Openings should be at least 750 mm from the internal (nearest) edge of a wall corner
- Openings formed by arches should have at least 450 mm of material above the crown

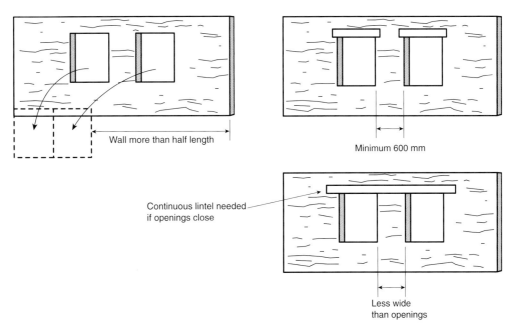

Figure 70 Simple rules for openings in rammed earth walls

6.4 Deformation

Elastic deformation
Elastic shortening of rammed earth under load may be calculated on the basis of the elastic modulus. Consideration should be given to the influence of moisture content on stiffness of walls that are likely to be loaded shortly after compaction.

Drying shrinkage
Where vertical drying shrinkage is considered to be significant to the structural design, material performance will usually be determined in advance from specific test data (Appendix A.3.4). Drying shrinkage may be used as a criterion for selection of suitable rammed earth materials; in this case drying shrinkage should normally be limited to less than 0.5%.

Movement joints
Movement joints are typically spaced 2.5–10 m apart, though walls may be built without joints when the expected deformations are sufficiently small. The location of movement joints in wall panels, such as in line with the edges of openings, should in general follow normal practice for masonry structures.

7 Maintenance and repair of rammed earth

Rammed earth is a durable construction material. Exposed to weathering and abrasion the surface conditions of rammed earth walls will often change during their service life. The nature and extent of these changes and maintenance procedures for rammed earth are presented here. Defects that may occur during construction of new walls are also outlined and repair methods discussed.

7.1 Weathering and deterioration

Weathering and deterioration of rammed earth walls is caused by many actions. Some of the most significant to new buildings include the following.

- **Exposure to wind-driven rainfall,** characterised by washing-out of fines over the surface and exposure of the gravel fraction (Figure 71). This process occurs most rapidly within the first year, and following the initial change will often stabilise to a much slower and relatively stable rate of weathering. Inclusion of fine gravel in the mix generally improves weather resistance. Slow changes in the surface through normal weathering require no further special measures. However, rapid localised loss of surface material and wall section will require repair and maintenance. Walls

(a) Before exposure   **(b)** After exposure

Figure 71 Surface weathering from rainfall

are most prone to rainfall damage immediately after compaction, as higher moisture levels weaken the material's strength and resistance.
- **Erosion due to concentrated water flow,** from overflowing drainpipes and blocked gutters. This may result in a much greater depth of erosion over a small area of the wall. Where plastic sheeting is used to protect walls during construction, gaps in the sheeting can lead to problems as the water flow is concentrated onto a small section of the wall (Figure 72). Sheeting may also encourage ponding of water on horizontal surfaces.

Figure 72 Concentrated rainwater flow damage

- **Splash-zone damage,** characterised by pitting of the wall surface along the base due to splash-back of falling rain and irrigation (see Figure 55). This is avoided by raising the wall base above the splash zone (225 mm) and limiting splash-back by providing extended eaves protection and soft ground surfaces, with hard surfaces inclined away from the wall and drains.
- **Abrasion damage** from construction plant and operations during building and general use on completion leading to gouging of the surface (Figure 73). Care must be taken to protect walls from other construction activities (Figure 74). Human contact, such as hand-rubbing, may also lead to light dusting (see Figure 82). Corners may be protected from damage by chamfering and raising the wall on an extended plinth.

Weathering and deterioration 87

Figure 74 Walls should be protected from other construction activities

Figure 73 Abrasion damage to vulnerable corners in a stabilised rammed earth wall

- **Localised water damage** in wet areas such as kitchens and bathrooms. Damage to walls in wet areas of a building is a perpetual concern, especially for those new to rammed earth construction. The reported incidence of damage is in fact quite low, however. In general, surfaces prone to prolonged wetting and possible flooding are protected by elevated plinths, screens and protective coatings where necessary.
- **Rising damp.** Damp rising from wet ground, often caused by bridging rather than failure of the damp-proofing barrier, may lead to deterioration through other means, such as rainfall weathering or abrasion, as persistent damp weakens the rammed earth.
- **Cracking.** Rammed earth has little tensile strength and so may crack because of material shrinkage, overload, ground movements (settlement, subsidence), poor detailing, vegetation (root) growth, and poor-quality site work (initial over-wetting, rapid drying-out, inadequate compaction). Cracking likely to impede airtightness, acoustic separation and fire barrier protection should normally not be tolerated in new rammed earth construction following drying. Cracking in rammed earth tends to be most extensive when clay contents and initial moisture levels are higher than

recommended. Vertical cracking generally occurs as a result of restrained material shrinkage. Horizontal cracking may coincide with formwork jointing, occur between compaction layers and follow 'over-compaction'.

7.2 Maintenance of rammed earth walls

All buildings require maintenance. Neglected buildings fall into disrepair as damp timbers rot, steel corrodes, concrete cracks, masonry and mortar joints spall, and protective coatings peel and fall away.

The nature and level of maintenance that rammed earth requires is specific to the characteristics of the material and use of stabilising additives, and may be higher than some other building materials; this is in part a consequence of using a building material that is readily recycled after use.

Natural earthen buildings will weather when exposed to rainfall. The design and maintenance of rammed earth buildings need to protect walls from excessive weathering to a level appropriate and acceptable to the expected design life of the building. The initial rate of weathering may be alarming to those unfamiliar with earthen buildings, but there are indeed many successful rammed earth buildings, in a wide variety of climates, that provide testimony to the ongoing success of this building technique.

As rammed earth buildings are increasingly built in areas where there is no significant historical tradition, possibly using sub-soils that also have largely not been used before, uncertainty over material performance is apparent. It is very difficult at present to predict the weathering of an untried rammed earth material. Performance depends on design and actual weathering over the life of the building, something increasingly difficult to predict as established weather patterns vary with climate change. Chemical and physical stability of sub-soil constituents should be considered. Externally exposed walls should normally be protected from the worst rigours of wind-driven rainfall, by screening if widespread weathering is to be avoided. Frost damage is generally less of a problem owing to the low moisture content of rammed earth.

Durability concerns are of course not limited to external exposed surfaces. Concerns about deterioration from surface abrasion have led rammed earth walls to be encased in glass (see Figure 12). Rubbing surfaces through manual contact may cause some loss of surface material, commonly known as dusting. This can be minimised by careful selection of a suitable material, but resistance may also be further enhanced by the use of protective coatings.

The aim of maintenance inspections is to detect early signs of deterioration to allow corrective action before more extensive works are required. Simple preventive actions, such as cleaning-out gutters and cutting-back vegetation, play an important role in all building maintenance. Regular inspections are essential to check for build-up of moisture and softening or breakdown of earthen materials. In many instances deterioration can be limited by routine maintenance work. Routine maintenance specific to rammed earth walls should include checking:

- Cracking of walls, which might suggest localised moisture increase (swelling), shrinkage, settlement, thermal shock, or overloading
- Water-borne erosion or spalling of material
- Integrity of protective coatings, including evidence of abrasive damage, cracking, erosion, peeling, spalling and separation
- Cleanliness and integrity of movement joints
- Integrity of damp-proofing and flashings
- Removal of plant growth

7.3 Defects in new construction

New rammed earth works should generally be free from cracks exceeding 3 mm in width, mechanical damage, loose sections of friable material, staining, and through tie holes. Many defects are a matter of aesthetic judgement (variations in colour and texture). However, other defects, such as boniness, may have significant influence on material integrity. Contract specifications should normally stipulate, possibly in terms of frequency per unit surface area and with reference to a test wall, agreed levels for:

- Colour variations (Figure 75). Caused by variation in materials, compaction, drying and formwork. An aesthetic concern that should be agreed in advance by reference to test wall and precedence study
- Textural variations (Figure 76). Caused by variations in materials, compaction and formwork. An aesthetic concern that should be agreed in advance by reference to test wall and precedence study
- Boniness (Figure 77). Caused by under-compaction and lack of fines within material at the surface. Boniness is both an aesthetic defect and a structural problem. In particular, sections of boniness should not be tolerated beneath bearings and other highly stressed areas of the wall (Figure 77(b))
- Surface level and flatness. Construction on completion of drying to be within specified tolerances (Section 5.10)

- Formwork patterning (Figure 78). Caused by gaps between formwork panels, holes and recesses in form faces and textural variations in formwork. An aesthetic concern that should be agreed in advance by reference to test wall and precedence study
- Surface cracking (Figure 79). Cracking should not impair aesthetic finish or airtightness of the wall. Surface cracking should generally not exceed 3 mm width and 75 mm length. More substantial cracking through the wall, likely to impair airtightness, should be filled or repaired
- Patching and repairs to surface defects (Figure 80). Plucking defects and through-bolt holes are commonly filled by plastering with similar material. However, because a different method of placement is used, a colour mismatch is often apparent. An aesthetic concern that should be agreed in advance by reference to test wall and precedence study
- Surface loss due to formwork adhesion (plucking) (Figure 81). Mostly an aesthetic concern that should be agreed in advance by reference to test wall and precedence study. May be patch repaired (see previous point)
- Surface friability (dusting) (Figure 82). Extent of surface friability should be agreed in advance by testing of materials (Appendix A.3.8) and by reference to test wall and precedence study
- Efflorescence/surface salt deposition (Figure 83). Salts present in sub-soils may occur on the surface after drying or following water damage. Light deposits may be brushed off without further problems. Heavy efflorescence may cause surface failure. In stabilised rammed earth, efflorescence is also caused by leaching out of calcium hydroxide

Colour and textural variations between subsequent layers are in general an intrinsic characteristic of rammed earth and usually cannot be considered a defect. Similarly, some limited cracking on drying and light boniness due to grading variations are to be expected.

To minimise their visual impact, repairs should use colour-matched materials. Joints, cracks and repairs can be further disguised by texturing or tooling fresh surfaces (see Figure 26). Repaired finishes may also be compared against an agreed test wall finish to ensure compliance with specifications.

Defects in new construction

Figure 75 Colour variation

Figure 76 Textural variation in a rammed earth panel

Figure 77(a) Boniness

Figure 77(b) Boniness beneath lintel support

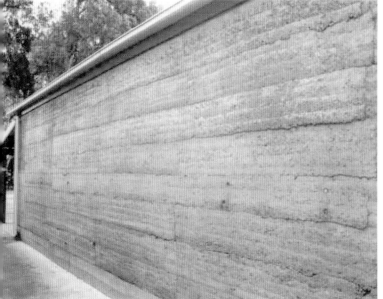

Figure 78 Formwork patterning

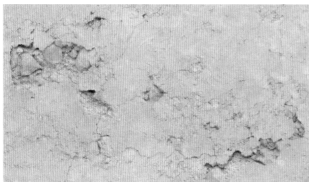

Figure 79 Surface cracking

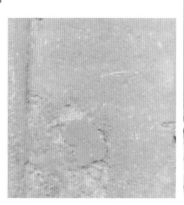

Figure 80 Patch repair   **Figure 81** Plucking damage

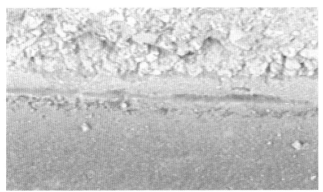

Figure 82 Surface dusting

Figure 83 Efflorescence in a stabilised rammed earth wall

7.4 Repairs to rammed earth

Repairs may be necessary as a result of defects during construction as well as because of damage occurring during the normal service life of buildings. Before undertaking repairs the cause of deterioration must of course be addressed. Repairs to historic and heritage buildings should follow the principles of architectural conservation.

Effective and routine maintenance regimes will minimise the need for repairs during the life of the building. Frequency of maintenance will depend on many factors. Initially maintenance inspections at least every 6 to 12 months are recommended, though these can be varied as building performance is established.

If possible, materials for repair should be taken from the same source as that originally used. Ideally a small quantity of original material should be set aside for such eventualities. Using the same material will help to ensure that repairs blend in, textural and colour match, with the original fabric. However, it is very difficult to undertake invisible patch repairs to rammed earth as the texture and colour often vary. To minimise visible impact of patch repairs the wall's surface may be tooled to change the texture and so hide the repair.

Materials used for repair must be compatible with the original fabric. Additives, such as cement, likely to impair vapour permeability should generally be avoided in repairing natural soils.

To hide small surface defects following construction remedial works such as bagging, patching, tooling, crack filling, and removal of salt deposits (efflorescence) may be required. Common defects in rammed earth, such as soft spots, boniness, surface cracking and formwork damage, can generally be repaired by surface patching and pointing. Deep repairs should be cut back and dowels may be used to ensure bond between the repair and original substrate. Light efflorescence should be easily removed by surface brushing.

Long-term success of patch repairs often relies on preparation of the original surface. Loose material is to be removed to provide a stable substrate. In general wall surfaces should be dry before applying the repair. Adhesion between deep patch repairs can be improved by using dowels embedded into the original wall material.

Patch repairs may be applied by trowel or float as a plaster. In such cases textural and colour match is likely to be poor. Repairs to wall bases, such as when erosion has underscored the wall, may be carried out by ramming in new material temporarily supported by formwork in front of the wall[6]. After placement the formwork is removed and the repair cut back to the line of the wall, ideally immediately following compaction whilst the material is still moist. The finished cut surface will generally not match the original, though after light weathering may eventually be indistinguishable from it.

Repairs may also be undertaken using pre-rammed or compacted blocks of material. This is particularly suitable for repairs above the base of walls. In general patch repairs are suitable for depths of up to 50 mm, and beyond this material should be cut back to receive a block or rammed repair. The joint between repair section and original material is likely to be clearly visible; where this not desired abrasion or tooling of the surface is an effective means of hiding such cracks. Further guidance on repair of earthen structures is available elsewhere[6].

Cracked and damaged renders should be repaired to ensure continued protection to the wall and limit likelihood of localised preferential weathering.

8 Future of rammed earth

Although the combined number of UK rammed earth and stabilised rammed earth structures is presently believed to be no more than several hundred, the last decade has seen a significant renewal of interest, driven primarily by the demands for more sustainable building. Over the past 25 years a few thousand stabilised rammed earth buildings have been built in Australia.

Recent applications of rammed earth in the UK have been varied, including visitors centres, a sports hall, a business park development, a children's nursery, a conference centre, as well as a prize-winning exhibition wall at the Chelsea Flower Show. New rammed earth projects currently under development include the Genesis Project at the Somerset College of Arts and Technology in Taunton (Figure 84), a 200-seat lecture theatre in the WISE Project at the Centre for Alternative Technology in Wales (Figure 85), and the Aykley Heads Development in County Durham.

(Architype)

Figure 84 Genesis Project, Somerset College of Arts and Technology

(Pat Borer & David Lea Architects)
Figure 85 WISE Project, Centre for Alternative Technology, Wales

There are considerable opportunities for the development of rammed earth in the UK in the foreseeable future. The technique offers a high-quality and sustainable building method suitable for a range of applications. Forms and finishes in rammed earth are wide and varied. Uses include loadbearing walls, non-loadbearing panels, external and internal situations, and single- and multi-storey construction. Though its main use has been through in-situ construction there are also significant future opportunities for off-site prefabrication of blocks, cladding panels and walls. No doubt the debate over the benefits and disadvantages of rammed earth and stabilised rammed earth (Appendix D) will continue as the two materials continue to develop and compete.

The lack of authoritative guidance has often been cited as a reason for the limited development and use of earth building in the UK. This has been the primary motivating factor for the production of this guide. As experience is gained the guide will develop too, perhaps one day into the production of a British or European code of practice for earth building.

Because of its poor thermal performance and susceptibility to water decay, one the greatest uses for rammed earth is likely to be internally, especially in combination with other construction materials, such as timber frame, providing structure, environmental regulation and beauty. However, innovative solutions for insulation performance of rammed earth walls, combined with weathering protection, will continue to develop providing greater flexibility for external wall construction. Prefabrication of panels and the greater use of permanent formwork are also likely to become more prevalent.

Though the project behind the development of this guide has taken significant steps forward in understanding material performance, many questions regarding rammed earth remain unanswered. Thermal performance, wider benefits of hygroscopic behaviour and breathing wall construction require further study in relation to rammed earth. One day a simple and reliable test to evaluate weathering resistance performance might even be developed.

Rammed earth has a long and successful history in the UK. Publication of this guide is another step in this process, one that we hope will become the platform for a wider renewal of interest and use of this beautiful and sustainable building material.

Appendix A

Physical properties of rammed earth

A.1 General

The level and extent of testing rammed earth materials depends on specific application and novelty of the material in use. Where a proven material is to be used then confidence in its qualities improves and the level of uncertainty and associated risk reduces.

For in-situ rammed earth, compliance tests are mostly undertaken on cylinders especially prepared for that purpose. In loadbearing applications it is usual to undertake soil classification, moisture–density (heavy manual compaction) testing, unconfined compressive strength and drying shrinkage assessment. Resistance to erosion and abrasion, flexural tensile and shear strength tests may also be carried out as necessary.

The rate at which materials are to be sampled for compliance testing should reflect the application and experience with the materials in use. Guidance for test procedures and the rate of testing are outlined in this appendix. Test conditions for specimens should as much as possible reflect the ambient or worst-case in-service conditions.

Pre-fabrication of rammed earth, in the form of either panels or blocks, enables some quality-control testing to be undertaken on the fabricated elements prior to their installation.

A.2 Compliance tests

Details and requirements for quality-control compliance testing of materials should be set out in the specification, as appropriate to the project, and agreed with all parties prior to commencement. The minimum performance specifications will vary with projects. Examples of appropriate specifications are given in Table A1. Suggested test methods for a range of parameters are set out in this appendix. Additional tests appropriate to the specification may also be used after mutual agreement amongst the parties involved.

Table A1 Typical minimum performance specifications for rammed earth

Parameter	Specification
Soil composition	Meet recommended and agreed specifications for grading, plasticity, shrinkage, chemical composition, mineralogy, colour, texture, organic matter content, and soluble salt contents
Minimum dry density	98% of heavy manual compaction test maximum dry density
Compaction moisture content	±1–2% of optimum moisture content
Unconfined compressive strength	1.0 N/mm^2 (general) 2.0 N/mm^2 (loadbearing)
Finish	Boniness, efflorescence, colour variation, etc, to be agreed in advance In general no cracks wider than 3 mm and longer than 75 mm
Erosion resistance	Erosion rate not greater than 1 mm/min
Surface abrasion	No general specification
Maximum drying shrinkage	Not greater than 0.5% (composite loadbearing) Not greater than 1.0% (other)

Physical property testing of in-situ placed rammed earth can be more problematic than other (prefabricated) materials and this needs to be addressed in any specification. Assessment of built walls is generally limited to aesthetic compliance. Surface hardness testing, using rebound hammers, has been used in the past, but results can prove variable and unreliable. Direct in-situ testing of density, such as sand-replacement or coring, is also problematic owing to low strength and relative slenderness of the elements. Indirect density tests, such as nuclear density testing used widely in other construction fields, may prove suitable for rammed earth in the future. Drying shrinkage measurements may be undertaken directly and reliably on walls, though the rate of drying needs to be considered; it may take many months for walls to dry out or reach relatively stable conditions. Therefore most compliance testing of rammed earth relies on preparation of cylinders for density and unconfined strength testing. Compliance testing, including moisture content and grading, can also be undertaken on samples of loose material during construction.

Extent of compliance testing will depend on the complexity of the project. The sampling rates and extent of testing need to be appropriate to the project. In some circumstances all

Test methods

assessment can be undertaken without laboratory testing, though in many cases some limited testing of materials will be necessary during initial selection of materials and for quality assurance during construction. Details of recommended field and laboratory testing procedures for materials and components are outlined in Table A2.

Table A2 Possible sampling rates for compliance testing during construction

Parameter	Sampling
Soil composition	Assessment during initial soil selection and delivery of each batch of material to site
Compaction moisture content	One representative sample per section of walling or one sample every 2 m^3
Unconfined compressive strength and minimum dry density	During initial material selection and 1 cylinder every 5–25 m^3 (depending on project specifications)
Finish	Test wall and on completion of construction and drying after specified period
Erosion resistance (optional)	During initial material selection
Surface abrasion (optional)	During initial material selection
Maximum drying shrinkage	During initial material selection and direct wall measurements after construction

A.3 Test methods

A.3.1 Test specimens

The test procedures set out below in general use small cylindrical or prism specimens. In preparation, particles greater than 19 mm are to be screened out. Inclusion of large particles in small specimens has a detrimental effect on performance. Where a significant proportion of material is screened out the specimen test may no longer be representative, achieving greater compressive strengths than those achieved with the larger size particles included. Therefore, for soils containing a significant proportion (20–30%) of material greater than 19 mm, testing of larger-size specimens, such as 300 mm cubes or 300 mm diameter cylinders, is recommended. Test procedures should otherwise, however, follow those set out below.

A.3.2 Moisture–density relationship

A.3.2.1 Background

The dry density of rammed earth material is primarily dependent on soil type, moisture content at compaction and compactive effort or energy. In order to achieve maximum density, it is important that the optimum moisture content, appropriate to the method of compaction, is used when ramming. However, the compactive energies of both pneumatic ramming and manual ramming are difficult to establish as they vary between operatives. The heavy manual compaction test and the 'drop' test are two commonly used procedures for establishing the optimum moisture content for rammed earth materials. These are both outlined below. The heavy manual compaction test should normally be undertaken in accordance with procedure outlined in BS 1377-4:1990[17].

A.3.2.2 Compaction test procedure

In this test a soil sample of known moisture content is compacted in a 1 litre (115 mm high × 105 mm diameter) cylindrical mould. Compaction is carried out in five layers of equal thickness by a 4.5 kg dropping weight falling 27 times on each layer from 450 mm. When the cylinder has been compacted to its full height within the standard mould, its wet weight is recorded to establish its bulk density. A sample of material is then taken for oven drying to establish the soil moisture content. At least five specimens at various moisture contents are prepared the same way and their bulk densities and moisture contents are recorded. After drying, the moisture contents and dry densities are calculated and plotted on a graph (of dry density and soil moisture content, see Figure 35). From the resultant curve, it is possible to determine the optimum moisture content for which the soil experiences its maximum dry density for the given compactive effort. The compactive energy of the heavy manual compaction test is widely believed to be lower than typical pneumatic works.

A.3.2.3 Drop test procedure

The 'drop test' is widely believed to provide a good approximation of the optimum moisture content. A ball of moist soil, approximately 40–50 mm diameter, is compacted by hand. The soil ball is dropped onto a hard flat surface from a height of approximately 1.5 m. When the soil is too dry the ball breaks into many pieces. When enough water has been added so that the ball breaks into only a few pieces, the soil is very close to its optimum moisture content. If the ball remains in one piece then the soil is too wet. The test is widely used as a means of controlling soil moisture content during construction. The drop test has a tendency to over-estimate optimum moisture contents, especially as the soil plasticity increases.

A.3.3 Unconfined compressive strength

A.3.3.1 Background

Compressive strength represents a basic quality-control measure for rammed earth, in the same way that cube testing does for concrete. The test to determine unconfined compressive strength of rammed earth is normally undertaken on cylinders prepared for that purpose. The cylinders are placed in a compression-testing device and loaded in uniform uniaxial compression until failure. Compressive strength is obtained from maximum applied loading and initial cross-sectional area.

Unconfined compressive strength is obtained by testing cylinders having a height to diameter ratio of 2. Alternatively the approximate unconfined strength of compaction test size cylinders (115 mm high × 105 mm diameter) may be established by factoring the test value by 0.7.

Cylinders are normally tested following drying to a stable moisture content under ambient conditions or curing for a specified period, such as 28 days.

A.3.3.2 Test procedure

Prepare five identical cylinders 200 mm high × 100 mm diameter or 300 mm high × 150 mm diameter. The material should be compacted at its optimum or, if different, the as-used moisture content, using the specified reproducible compactive effort. Compactive effort is usually the heavy manual compaction test, using a 4.5 kg hammer, though pneumatic ramming may also be specified. The maximum aggregate size should not exceed one-sixth of the cylinder diameter. Where more than 20–30% of material is screened out this change in grading may have a significant influence on recorded strength, and a larger more representative specimen should be used (see A.3.1).

After compaction, the cylinders should be stored for drying or curing. Rammed earth specimens may be dried at 15–20 °C and 40–60% RH until there is no further loss of moisture. Specimens containing cement or another stabilising additive should normally be cured under appropriate conditions for the specified period (eg 28 days) and then tested after laboratory oven drying.

Cylinders should be capped after drying and before testing to provide two opposing parallel and flat surfaces. The capping material may be dental plaster or similar material and should not exceed 5 mm in thickness at either end. Before capping, measure and weigh each cylinder to establish material bulk density.

Ensure that the end surfaces are clean and place the cylinder between the compression testing platens. The load is to be applied without shock and increased continuously until failure.

In strain-controlled devices the moving head should travel at a rate of 1.0 mm/m strain per minute. In load-controlled devices, the load should be applied at a constant rate equivalent to a specimen stress of 0.2 N/mm² per minute. Record the maximum load and mode of failure. If required, set aside representative samples of material from each test cylinder to determine moisture content at testing.

For cylinders of height to diameter ratio of 2, the unconfined compressive strength of each specimen is equal to its maximum load divided by initial cross-sectional area. The characteristic unconfined compressive strength is given by:

$$f_c = f_a - 1.65\sigma_{n-1}$$

where: f_c = characteristic unconfined compressive strength of test sample
f_a = average unconfined compressive strength of test sample
σ_{n-1} = standard deviation of test sample

The test report should include:
- Specimen dimensions
- Moisture content at compaction
- Compactive effort
- Percentage of material screened out during preparation
- Drying conditions
- Capping material
- Moisture content at testing
- Cylinder dry density
- Average and characteristic unconfined compressive strength

A.3.4 Drying shrinkage
A.3.4.1 Background
The test is a measure of how much rammed earth materials shrink linearly on drying following compaction. Atmospheric conditions (temperature, relative humidity) will determine both the rate and final drying shrinkage of rammed earth. Drying shrinkage tests are recommended where the differential shrinkage of rammed earth may have significant influence on loadbearing walls. Tests may be undertaken on cylinders prepared for unconfined compressive strength testing (A.3.3). Where more than 20–30% of material is screened out, the proportionally greater fines content of the cylinders may overestimate actual material shrinkage. Proportion of materials screened out in preparation should be reported.

A.3.4.2 Test procedure

Five cylindrical specimens should be prepared for compressive strength testing. Following compaction the cylinders should be allowed to dry out under controlled temperature and RH conditions. Recommended test conditions should remain constant and within the ranges 15–20 °C and 40–60% RH, though these may be varied to suit specific applications.

Determination of linear shrinkage relies on measurement of total or relative changes in length of the cylindrical samples. Measurements may be made directly using total length, before and after drying, or more precisely using surface-mounted strain devices such as a DEMEC gauge. Initial measurements should be taken immediately following compaction and demoulding of the cylinders and thereafter periodically during drying. Shrinkage measurements cease when they no longer change with time and when the cylinder mass remains constant. Change in moisture content may be determined from an additional specimen prepared for that purpose and subject to the same test conditions.

Cylinder shrinkage is expressed as the ratio of change in length to original datum length. The test report should include:
- Average drying shrinkage of the cylinders
- Compactive effort
- Proportion of material screened out
- The initial (compaction) moisture content
- Final moisture content
- Period of time for shrinkage to occur
- Test conditions during drying
- Method of measuring shrinkage

A.3.5 Flexural tensile strength

There is no recognised test procedure for flexural tensile or bending strength testing of rammed earth. In most applications knowledge of flexural strength is not required. However, when considered necessary, tests should normally seek to determine flexural strength perpendicular to the horizontal compaction layers.

The test specimen should allow application of flexural tensile stress in bending perpendicular to compaction layer bedding. The specimen should be of sufficient size to allow a number of identical compaction layers to be tested under conditions of increasing uniform bending moment. The specimens should be prepared under the heavy manual compaction test or other specified and reproducible compactive effort. After compaction the specimens should be dried or cured and dried under agreed conditions, such as those specified for compressive strength testing (A.3.3). If more than 20–30% of material is screened out this change in grading may have a significant influence on recorded strength, and a larger more representative specimen should be used.

Flexural tensile stress may be applied to test specimens either as a four-point load beam test or using a bond wrench[22] or similar device. The influence of self weight should be taken into account when determining flexural tensile strength. The load should be applied to the test specimens without shock at a uniform rate until failure. Testing should normally be complete within 60–120 seconds of applying the load. The flexural tensile strength of each specimen should be calculated based on gross cross-sectional properties.

The test report should normally include:
- Size of test specimen
- Compactive effort details
- Proportion of material screened out
- Details of soil preparation
- Rammed earth moisture content at compaction and testing
- Curing and drying conditions
- Details of test set-up, including loading rate
- Individual specimen results
- Average and characteristic flexural tensile strength

A.3.6 Shear strength

Knowledge of shear strength may be required in design to assess racking shear resistance of (non-loadbearing) rammed earth walls. Shear strength may be determined using large-scale (ie 300 mm × 300 mm) shear box testing or direct tests on prototype walls (Figure A1). Preparation of materials and specimens should replicate likely site conditions.

The test report for shear testing will normally include:
- Details of specimen preparation, including materials
- Details of test set-up and rate of loading
- Shear strength
- Coefficient of friction (when testing under increasing normal stress such as the shear box)

A.3.7 Accelerated erosion test

A.3.7.1 Background

At present there is no simple or reliable test to predict the erosion resistance of rammed earth materials. Test methodologies including water-spray, water-drip or wet–dry tests may be used to compare relative performance of different materials. The accelerated erosion test, as set out in *The Australian earth building handbook*[3], is recommended to determine relative erosion resistance of rammed earth panels. Specimens are subject to a continuous jet of water spray for

Figure A1 Shear testing of rammed earth wall panel

60 minutes or until the water has completely penetrated the specimen. Performance, in terms of erosion rate, is determined on the basis of pitting depth or time taken to penetrate the specimen completely.

A.3.7.2 Test procedure

Normally three identical rammed earth specimens should be prepared for the erosion test. The specimens should measure 300 mm × 300 mm (test face area) and normally be at least 100 mm thick. Thickness should be adequate to allow a representative material grading in testing. The specimens shall be compacted using the heavy manual compaction procedure or another specified and reproducible compactive effort. The rammed earth layers shall be compacted in equal layers across the length and thickness of the specimen. Each specimen will be allowed to dry out and protective coatings applied as required in preparation for testing.

Appendix A: Physical properties of rammed earth

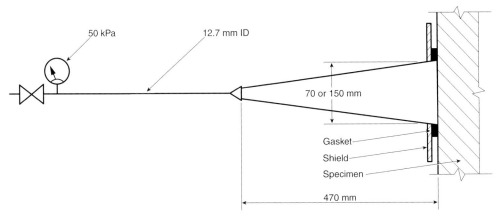

Figure A2 Spray erosion test

The specialised apparatus required for the test is shown in Figure A2. The test includes a 50 mm spray nozzle mounted in front of the specimen, a water pump and pressure gauge, settlement tank, filtration screen to remove particulate matter, and mountings for the specimen, including a screen and gasket.

Specimens are to be mounted in the test rig in the same orientation as that intended for the wall construction, ie with the compaction layers horizontal. The screen should be positioned such that a limited area of one face of the specimen is exposed to the spray. Each specimen is subjected to the water spray for 60 minutes or until the specimen has completely eroded through.

At the end of testing the depth of erosion is measured to the nearest millimetre using a 10 mm diameter flat-ended rod. The maximum depth measured for each specimen is taken as the rate of erosion (mm/hour). Where the specimen fails in less than 60 minutes the rate of erosion is determined from specimen thickness divided by the time (in hours) at which the test was stopped. The average test performance for the sample of three specimens is to be reported.

A.3.8 Abrasion resistance

A.3.8.1 Background

At present there is no universal recognised test for expression of abrasion resistance for earthen materials. Laboratory tests generally comprise wire-brush abrasion of surfaces under a specified pressure and measurement of material lost. Abrasion testing may be useful for comparing the performance of different materials or protective coatings, but is unlikely to provide a reliable predictive indication of future performance. Variations in surface finish of rammed earth, such as boniness, can lead to significant varying performance of an individual test panel. A simple test procedure, developed for compressed earth blocks, is outlined below[23]. Other methods, such as those developed by Minke[9], might also be used.

A.3.8.2 Test procedure

The basic objective of the test is to measure the resistance to abrasive damage of a rammed earth surface. The rammed earth test panel is subject to abrasive action of a metal wire brush, at constant vertical pressure, for a given number of cycles. The rammed earth panel surface should be prepared to be representative of the proposed or actual wall construction. The specimen should measure 300 mm × 300 mm × up to 100 mm thick. The abrasion resistance is measured as loss of material (mass) over the test surface area.

The test uses a standard wire brush. To maintain a constant vertical pressure a mass of 3 kg is fixed to the brush (Figure A3). In preparation for testing the specimen(s) should be dried to constant mass under specified conditions. The specimen should be capable of being laid flat for testing. The specimen is initially weighed. The test face is then subject to wire-brush abrasion, which comprises 60 complete backward and forward cycles in 1 minute. No further vertical pressure should be applied during

Appendix A: Physical properties of rammed earth

Figure A3 Abrasion test

testing. The width of the brushed area should not exceed the width of the brush by more than 2 mm. The brushing should be along the full 300 mm length of the panel and throughout brushing at least half the surface of the brush should remain in contact with the test surface. On completion of brushing lightly remove any remaining loose material and re-weigh the panel. Report the abrasion resistance of the test panel as the area of the brushed surface divided by the mass reduction caused by brushing (m^2/kg). The direction of brushing may be parallel or perpendicular to the direction of the compaction layers. Normally more than one test may be undertaken on both sides of each specimen.

The test report will normally include:
- Details of specimen preparation, including protective coatings
- Specimen density and moisture content
- Average abrasion resistance of each specimen
- Average abrasion resistance of sample
- Direction of brushing relative to compaction layers

Appendix B

Specification for rammed earth works

This appendix sets out suggested details for inclusion in a specification for rammed earth works. It is not a complete specification, but concentrates on those aspects specific to rammed earth works. More general items, such as responsibility for documentation management, have not been included. The specification is accompanied by some commentary presented in italics.

1 **Scope of rammed earth works to be undertaken**

   All rammed earth works shall be constructed with appropriate care, site control and supervision so that the minimum design requirements of this specification are met.

   Construction is to include all rammed earth walls from and including the damp-proof course to the top of the rammed earth wall, including installation of holding-down anchors or ties and wall plates.

   This specification may choose to exclude the damp-proof course if installed within a plinth, and fixings such as wall plates and holding-down anchors or ties.

2 **Submissions to be made**

   The rammed earth mix proposal, including details of soil particle size distribution (determined in accordance with BS 1377-2:1990[16]), soil plasticity (BS 1377-2:1990[16]), soil mineralogical analysis, heavy manual compaction test (BS 1377-4:1990[17]), drying shrinkage (Appendix A.3.4)*, unconfined compressive strength (Appendix A.3.3), spray erosion resistance (Appendix A.3.7), and abrasion resistance (Appendix A.3.8). Details of test procedures and results to be submitted at least 1 month prior to rammed earth works starting on site.

   Spray erosion resistance and abrasion resistance may be considered optional items for inclusion in a specification.

---

* Appendix number references are to the appendices in this guide *Rammed earth: design and construction guidelines*.

3 Tests to be undertaken by the rammed earth contractor during construction

These tests are undertaken to demonstrate compliance of the materials with the specification requirements. These requirements will have been established during initial testing of materials prior to construction.

Test	Frequency
Particle size distribution	As noted in Section 2 and when material or source is varied
Unconfined compressive strength	*To be specified by engineer*
Moisture content	*To be specified by engineer*
In-situ bulk density	*To be specified by engineer*

In-situ tests, using sand replacement or another accepted method, may be undertaken to check achieved dry density, though excavation-based methods are very difficult to do without risk of damage to rammed earth works. Indirect methods for assessing quality of works include surface rebound hardness testing such as the Schmidt hammer.

4 Notice to inspect

The contractor shall give at least 48 hours' notice to allow the opportunity to inspect the following:
(a) Damp-proof course installation
(b) Construction joints
(c) Flashings installation

Joints in damp-proofing membranes must be lapped or sealed against moisture penetration. Damp-proofing and flashings should not be breached during construction. Any breach or damage shall be made good. Where necessary to prevent moisture ingress, flashings must project at least 25 mm from the face of the wall.

5 In-situ rammed earth mix and supply

Details of the rammed earth mix design(s) must be submitted, by the supplier, to allow a minimum of 1 month for approval.

Materials on site should be handled and stored in a way that ensures that their performance is not impaired. Any materials that have deteriorated sufficiently to impair their performance should be rejected.

5.1 Compressive strength

The mix shall achieve a minimum characteristic unconfined cylinder dry compressive strength of ❏ N/mm^2 *(strength to be specified by engineer)*. Cylinders shall be prepared and tested in accordance with Appendix A.3.3.

5.2 Mix constituents

The rammed earth will contain no additives or admixtures *(unless agreed by the engineer)*.

The soil grading curve shall meet following limits:

Sand + gravel content	45 to 80% (by mass)
Silt content	10 to 30% (by mass)
Clay content	5 to 20% (by mass)

The maximum aggregate size shall not exceed 6 mm (or 10 mm, 20 mm, 25 mm, 38 mm or 50 mm) *(as specified by the engineer)*.

The material shall have a combined organic matter, soluble salt and other deleterious matter content not exceeding 1–2% *(as specified by the engineer)*.

The material shall be properly mixed to provide an even and consistent material. Mix proportions of all materials should comply with those specified. Methods used to measure materials should ensure that specified proportions are controlled. Mixing should ensure that all ingredients are evenly distributed throughout the mixture.

5.3 Moisture content

The moisture content for the rammed earth at time of compaction shall be within ±1–2% *(as specified by the engineer)* of the agreed optimum moisture content for compaction and handling. A heavy manual compaction test shall be undertaken to determine the optimum moisture content at least 1 month before the start of work. The moisture content shall normally be regularly checked *(ie by drop test, micro-wave oven or carbide meter)* to ensure a consistent and uniform material throughout construction.

5.4 Drying shrinkage
The drying shrinkage of the rammed earth, measured in accordance with Appendix A.3.4, shall not exceed 0.1% (or 0.2%, 0.5% or 1%) *(as specified by the engineer)*.

5.5 Material supply
In contracts where some or all of the materials are imported.
The name and address of supply depot(s) is to be submitted before delivery of the material. All delivery notes are to be retained for inspection.

6 Testing and certification of materials

6.1 Recording
Representative location of specimens prepared and submitted for testing shall be recorded.

6.2 Test laboratory
All material testing shall be undertaken by an approved laboratory.

6.3 Sampling rates
The rate of materials sampling for testing of the following parameters shall be specified by the engineer: unconfined compressive strength; moisture content at compaction; in-situ density test; drying shrinkage; grading.

6.4 Compression test specimens
Cylinders shall be prepared and tested in accordance with Appendix A.3.3. Cylinders that break during preparation or fail to meet specified requirements shall be retained for inspection.

6.5 Failures
If a rammed earth sample fails to achieve specified requirements following construction or testing, inform the engineer and submit:
- confirmation of the validity of the test results, and
- proposals for further tests, or
- proposals for rectification.

Obtain approval of all such evidence and proposals before proceeding.

7 Placement and compaction

7.1 Preparation
At time of placement, all surfaces, including shuttering forms which rammed earth is to be compacted against, shall be clean and free of debris and excess water.

7.2 Transportation
During transport of rammed earth from mixing area to site for compaction, avoid contamination, segregation, loss of ingredients and excessive evaporation. Cover rammed earth during heavy rain.

7.3 Placement
Place loose prepared material in the formwork in courses of even and controlled depth. The maximum depth of loose material is not to exceed 150 mm.

Cold joints between lifts should be protected from excessive drying and scarified before proceeding with further works.

Record the time, date and location of all rammed earth works. During compaction ensure that the temperature of rammed earth does not exceed 30 °C or fall below 5 °C. Do not place against frozen or frost-covered surfaces. The rate of construction should be regulated to minimise risk of deformation or instability.

7.4 Compaction
Rammed earth material is to be fully and properly compacted, taking especial care around inserts, formwork corners and at joints. The rammed earth should achieve a minimum dry density of not less than 98% of the heavy manual compaction test maximum dry density.

8 Drying and protection

8.1 Drying
Protect walls in cold weather from frost damage throughout the drying period. Detailed records of the location and timing of compaction of individual batches, removal of shuttering and removal of coverings are to be maintained on site for inspection.

8.2 Protection
Protect walls from direct rainfall, splash-back and runoff by means of roof protection and surface coatings as necessary. Protect walls from abrasion, other physical damage arising from construction works, thermal shock, impact, overloading, movement and vibration. Take care to avoid uneven drying of walls where one face is in direct sunlight and the other shaded, which may lead to leaning or bowing of walls.

Waterproof sheeting used to protect walls from rain damage should be held clear of the surface to allow air circulation and to allow drying out of the rammed earth walls.

9 Formwork

9.1 Construction

Construct formwork accurately and robustly with adequate supports to produce finished rammed earth to the specified dimensions. Formed surfaces must be free from twist and bow (other than any required cambers), with all intersections, lines and angles being square, plumb and true.

Construct formwork, including joints in form linings and between forms and completed work, sealing joints where necessary. Secure formwork tight against adjacent rammed earth to prevent formations of steps.

Confirm positions and details to ensure that alterations to and decisions about the size and location of inserts, holes and chases are not made without the knowledge and approval of the engineer. Fix inserts or box out as required in correct positions before placing rammed earth. Form all holes and chases.

No metal part or device for securing forms, such as ties, is to remain within the completed rammed earth.

No release agents *(unless otherwise specified)* are to be used on the formwork faces.

9.2 Striking of formwork

Strike formwork without disturbing, damaging or overloading the structure, and without disturbing props. Notwithstanding other clauses in this specification and any checking or approvals by the engineer, the responsibility for safe removal of any part of the formwork and any supports without damaging the structure, rests with the rammed earth contractor.

9.3 Formed finishes

All finished and visible surfaces on rammed earth walls should be free from cracks exceeding 3 mm in width, mechanical damage, sections of loose friable material (soft spots), staining, and open bolt-holes.

Wall finishes shall meet minimum agreed variations with respect to colour, texture, boniness, flatness, formwork patterning, allowable cracking, extent of patching and repairs, and the number and style of cold joints. Any defect should not impair either the form or the function of the wall. All repairs should be undertaken using colour-matched similar materials. The visible surface standard of the completed wall, including all repairs, should be measured against that of an agreed test wall or other agreed reference finish.

Specification for rammed earth works

The completed work is to produce an even finish with panels arranged in a regular pattern as a feature of the surface. Abrupt irregularities are not to be greater than 2 mm. Gradual irregularities, expressed as maximum permissible deviation from a 1 m straight edge, are to be not greater than 5 mm. Formwork tie holes to be in an approved regular pattern, filled with matching earth.

Construction tolerances

Horizontal position of any rammed earth element specified or shown in plan at its base or at each storey level	±10 mm
Deviation within a storey from a vertical line through the base of the member	±10 mm per 3 m of height
Deviation from vertical in total height of building (from base)	±15 mm per 7 m of height
Deviation (bow) from line in plan in any length up to 10 m	±10 mm per 5 m of length
Deviation from vertical at surface against which joinery is to be fitted	±10 mm
Deviation from design wall thickness	±10 mm
Position of individual rammed earth formwork panels	± 5 mm

The walls are to be inspected for compliance with this specification ❑ weeks *(normally between 4 and 12 weeks)* after construction, to allow adequate drying of the walls, and following installation of protective coatings.

10 Movement joints in rammed earth

Movement joints likely to open as the material shrinks should be constructed in such a way as to prevent ingress of moisture and to remain airtight. Repairs may be undertaken using earthen materials or suitable proprietary fillers.

Movement joints likely to close (thermal expansion or articulation) should be clean and free from any hard or incompressible material for the full width and depth of the joint before any joint-filling material is inserted.

All joints are to be accurately located, straight and well aligned, truly vertical or horizontal or parallel with the setting out lines of the building.

The contractor shall prepare a detailed layout of construction joints for approval by the engineer prior to any rammed earth being placed. Joints are to be formed accurately. Modifications to joint design are to be approved by the engineer before proceeding.

Formed joints are to be constructed using rigid forms or stop ends designed to avoid temporary bending or displacement.

11 Worked finishes in rammed earth

Tops of rammed earth walls are to be constructed sufficiently flat and level *(±20 mm)* to receive a timber wall plate bedded in mortar, a reinforced concrete ring beam or other specified fixture.

12 Fixings, fittings and embedded items in rammed earth

Fixings are to be located a minimum of 150 mm from the edge of any wall.

Provision should be made as the work proceeds for all partitions, straps, beams, trusses, plates, and the like which are to be built or keyed into the wall, so as to minimise subsequent cutting or chasing of the wall. In rammed earth it is often easier to fix items such as straps and ties after wall construction is completed. Proprietary fixing types and methods of installation to be specified here.

13 Completion and maintenance

Contractor is to provide a maintenance manual describing care and maintenance of the walls.

Details on types and methods of application of protective coatings, renders and other finishes to be provided here.

Appendix C

Structural wall design

When a more rigorous approach to that set out in Chapter 6 for checking the structural resistance of rammed earth walls is required, a procedure based on loadbearing masonry design may be used[3,24]. The methodology is based upon a limit state philosophy in which characteristic compressive strengths and factored design loads (actions) are used. Other suitable, recognised and accepted structural design approaches may also be employed.

C.1 Material partial safety factor (γ_m)

A partial safety factor is applied to material property design values to account for variations in materials and quality of work. For example, material properties are often determined using small specimens prepared under laboratory conditions, and so do not include features such as boniness. Criteria influencing the partial safety factor value, together with outline values, are summarised in Table C1. Selection of the partial safety factor is at the engineer's discretion. There are insufficient data to make more specific recommendations. The recommended value for partial safety factor varies between 3 and 6, though design engineers may select alternative values as they consider appropriate. Designers should also consider the possible consequences of failure and likelihood for accidental damage.

Table C1 Values for material partial safety factor (γ_m)

Suggested criteria	(γ_m)
Works carried out by experienced specialist contractor; tried and tested materials; materials from consistent supply or mix; materials tested fully in accordance with provisions of Appendices A and B; full programme of compliance testing during construction; materials well within recommended limits of suitability criteria; material property test results demonstrate consistent repeatable performance	3.0–4.0
Works carried out by general contractor under supervision; untried material with limited laboratory test data; full programme of compliance testing during construction; materials within recommended limits of suitability criteria	4.0–5.0
Works carried out by inexperienced labour under some supervision; untried natural or quarry waste material with limited test data; limited programme of compliance testing; materials marginally comply with recommended limits of suitability criteria; material property test results show some inconsistency	5.0–6.0

C.2 Design for combined compression and bending

The vertical forces and moments are combined at the top and bottom of the wall by regarding the vertical force as acting at a statically equivalent eccentricity (e) at each end. In designing the wall the most unfavourable combination of imposed actions should be considered.

Compressive strength is a function of wall slenderness ratio (S_r), load eccentricity (e), material compressive strength (f_c), and wall section dimensions breadth (b) and thickness (t). Slenderness ratio (S_r) is given by:

$$S_r = h_{ef}/t$$

where the effective height is a function of lateral restraints at the base and top of the wall as follows:

- h = clear wall height between restraints
- h_{ef} = $0.75h$ for a wall laterally supported and rotationally restrained both top and bottom
- h_{ef} = $0.85h$ for a wall laterally supported both top and bottom and rotationally restrained along at least one of these
- h_{ef} = $1.00h$ for a wall laterally supported but rotationally free both top and bottom
- h_{ef} = $2.00h$ for a wall laterally supported and rotationally restrained only along its bottom edge

For sufficient compressive strength the wall must satisfy the following basic requirement:

$$N_d \leq (\Phi f_c\, b\, t)/\gamma_m$$

where:
- N_d = design compressive force
- Φ = capacity reduction factor, dependent on slenderness ratio and load eccentricity, as defined in Table C2
- γ_m = material partial safety factor
- f_c = unconfined material compressive strength

Table C2 Slenderness and eccentricity reduction factor (Φ)

Slenderness ratio* (S_r)	Reduction factor (Φ) Ratio of maximum eccentricity to thickness (e_{max}/t):			
	≤ 0.05	0.10	0.20	0.30
6	1.00	0.78	0.56	0.32
8	0.94	0.73	0.54	0.29
10	0.88	0.67	0.49	0.25
12	0.82	0.62	0.45	0.22
14	0.76	0.56	0.40	0.18
16	0.70	0.51	0.35	0.15
18	0.64	0.45	0.31	0.11

* Slenderness ratios above 12 (shaded) are greater than recommended for general construction.

C.3 Design for concentrated compression loads

An increase in compressive capacity of up to 50% is permitted in zones under concentrated loads. Though concentrated loads are assumed to disperse through the rammed earth at an angle of 45° from the perimeter of the bearing area of the load, the dispersion cannot extend:

(a) into the dispersion zone of an adjacent concentrated load, or
(b) beyond the physical end of the wall or across movement joints.

The wall is designed to satisfy the following condition for each cross-section within the zone of dispersion of the concentrated load:

$$N_d \leq (\Phi_b f_c A_b)/\gamma_m$$

where: N_d = design compressive force, including the concentrated load and portion of any other compressive forces acting on the cross-section under consideration
Φ_b = concentrated bearing factor
A_b = area beneath bearing taking account of load distribution
γ_m = material partial safety factor
f_c = unconfined material compressive strength

Appendix C: Structural wall design

The value of Φ_b is taken as follows:

(a) For cross-sections at a distance greater than $0.25h$ below the level of the bearing:

$$\Phi_b = 1.00$$

(b) For cross-sections at a distance within $0.25h$ below the level of the bearing of the concentrated load on the member:

$$\Phi_b = [0.55(1 + 0.5a_1/L)]/(A_{ds}/A_{de})^{0.33} \text{ or}$$

$$\Phi_b = 1.50 + (a_1/L)$$

whichever is less, but Φ_b not less than 1.00, or greater than 1.50,

where: A_{ds} = bearing or dispersion area of the concentrated load at the design cross-section under consideration
A_{de} = effective area of dispersion of the concentrated load at mid-height (see Figure C1)
a_1 = distance from the end of the wall to the nearest end of the bearing area
L = clear length of the wall
L_e = effective length of load dispersal at mid-height of the wall
t = section thickness

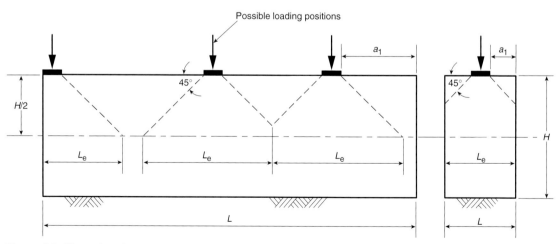

Figure C1 Dispersion of concentrated loads ($A_{de} = L_e t$)

C.4 Design for out-of-plane flexural loads

The design of a rammed earth wall to withstand vertical bending from actions of a short-term transient nature, which include out-of-plane wind loads or similar forces, should either satisfy requirements of combined bending and compression or satisfy:

$$M_d \leq [(f_t/\gamma_m) + f_d] Z$$

where: M_d = vertical design bending moment, including bending action from load eccentricities or bending moments applied at the ends of the wall
f_t = flexural tensile strength of rammed earth
f_d = design compressive stress at the cross-section
γ_m = material partial safety factor
Z = section modulus

C.5 Design for shear

The design of a rammed earth wall cross-section to withstand shear forces is undertaken to satisfy the following relationship under each combination of simultaneously acting design shear force (V_d) and minimum design compressive stress (f_d):

$$V_d \leq (v_o/\gamma_m + \mu f_d) A_v$$

where: v_o = basic shear strength of rammed earth (determined by test – Appendix A.3.6)
A_v = area of cross-section resisting shear
μ = shear factor
γ_m = material partial safety factor

Appendix D

Stabilised rammed earth

D.1 What is stabilised rammed earth?

Stabilised rammed earth is a specific form of rammed earth construction which uses sub-soils combined with stabilising agents, in particular ordinary Portland cement, to improve the material's physical characteristics. Cement stabilisation of rammed earth soils has become common accepted practice in Australia. The addition of cement significantly improves wet compressive strength resistance, resistance to water-borne deterioration, and general durability and robustness. Its use is considered to reduce uncertainty and risk involved in the use of novel materials. However, these advantages need to be very carefully weighed against the environmental impact of cement production, which accounts for some 8% of total carbon dioxide

(Matthew Hall; 2003)

Figure D1 Brimington Bowls Club Pavilion, Chesterfield, stabilised rammed earth

(Earth Structures Pty Ltd; Bill Swaney; 2001)

Figure D2 Stabilised rammed earth stables, Ashley, Northamptonshire emissions worldwide[25]. Cement also fundamentally alters the structure of rammed earth materials, impeding ability to be recycled on end use. Portland cement is added during soil preparation and mixing on site at between 4 and 12% by mass.

The construction process for stabilised rammed earth is identical to that for rammed earth. Improvement in physical characteristics allows a degree of greater flexibility in design and application. During the past 30 years considerable experience and expertise in stabilised rammed earth has been gained in Australia and parts of the United States of America. To date, stabilised rammed earth building has proven to be a durable form of construction and has been used in a great variety of projects. This experience has also been imported to the UK in recent years (Figures 22, 23, D1 and D2). Recent innovations include the inclusion of sandwich wall rigid foam insulation inside stabilised rammed earth walls and stabilised rammed earth cladding panels.

Cement stabilisation is most suited to soils with relatively lower clay contents than those used for (natural) rammed earth. Cement paste binds the soil aggregates together in a rigid structure similar to the way it does in concrete, working most efficiently for soils with low clay contents. Clay content must be limited to ensure adequate durability and dimensional stability. As certain clays are more reactive than others, clay type as well as the overall amount is important. Total clay content is readily obtained from the sedimentation test, whereas clay type analysis requires more rigorous mineralogical testing. Organic matter content should normally not exceed 2% as it is harmful to cement hydration. Soils with soluble salts (sulfates of sodium, potassium, magnesium, calcium) in sufficient quantities (2–3%) to impair strength or durability (found by trial testing) are unsuitable. Generally soils for cement stabilisation should meet following criteria:

Sand + fine gravel content	45 to 80%
Silt content	15 to 30%
Clay content	Up to 20%
Plasticity index	2 to 22 (liquid limit <40)

CRATerre has produced guidelines on soil suitability for cement stabilisation based on soil plasticity[4].

D.2 Advantages and drawbacks of stabilised rammed earth

D.2.1 Advantages

- **Improved durability**

  The addition of ordinary Portland cement to sub-soil improves the material's resistance to water. Stabilised rammed earth may readily be immersed in water for prolonged periods without loss of integrity. For this reason it may be used below damp-proof course level as a foundation for rammed earth. Resistance to rainfall erosion and abrasive damage can be significantly enhanced. Cement paste binds together fines content to provide surfaces that do not readily dust when abraded. The ability to withstand higher levels of moisture without loss of integrity offers the opportunity for inclusion of impermeable materials such as rigid foam sandwich insulation. The enhancement in durability depends on the use of appropriate sub-soils and level of cement stabilisation.

- **Improved strength**

  Stabilisation with cement improves the mechanical strength of sub-soils. The improvement is most pronounced when materials are wet, though there is also often improvement in dry strength as well. Cement-stabilised sub-soils can readily develop compressive strengths in excess of 10 N/mm^2. The enhancement in strength depends on the use of appropriate sub-soils and level of cement stabilisation. Compressive strength improvement follows a linear relationship with increasing cement content[4]. Enhanced strength allows the opportunity for thinner walls, although construction requirements often prevent this, and resistance to higher loads from greater floor loadings and taller buildings. Higher strengths, and generally associated higher material stiffness, provide the opportunity for reinforcing rammed earth with materials such as steel. Greater material resistance can also remove the need for wall plates and ring beams.

- **Reducing perceived risk**

  Builders, architects, engineers and clients are accustomed to using materials that often have much greater strength and durability required for many building applications. The inclusion of cement produces a material more akin to concrete and other familiar cementitious products. These improvements in strength and durability offered by using stabilised rammed earth lower the perceived risk of material performance. Stabilised rammed earth can be used as a direct replacement of other materials, such as concrete block and fired brick, without significant variation to building design.

D.2.2 Drawbacks

- **Environmental impact of cement stabilisation**
  Cement production is unquestionably a major contributor to manufactured CO_2 emissions, a major greenhouse gas. There is approximately 0.8–1.0 tonne of CO_2 embodied in every tonne of cement produced. As a comparison the environmental impact of heavy goods vehicle road haulage is approximately 2.5–4 kg of CO_2 per tonne (gross vehicle mass) for every 100 km travelled (figures from UK Department of Transport). The inclusion of pozzolans, such as ground granulated blastfurnace slag (ggbs), an industrial by-product, can reduce cement content. Stabilised rammed earth typically contains 6% cement (by dry mass), compared with concrete blocks that may contain around 4–10% (typically 5–6%) cement (by dry mass) (figures from UK Concrete Block Association). Consequently the material in a 300 mm thick internal stabilised rammed earth wall is likely to contain significantly greater amounts of cement than an equivalent single skin 100 mm thick dense concrete block wall. Stabilised rammed earth walls offer a finished product, not requiring plaster or render coats, however. Cement stabilisation may remove the need for reinforced concrete ring beams and application of protective coatings, allow reduced wall thickness, and reduce the need for extended eaves protection and maintenance. However, these benefits should be carefully balanced, over the expected life-time of the building, against the environmental impact of using cement.

- **Loss of material quality**
  Combining cement with soil aggregates too closely resembles concrete for many people attracted to earthen architecture. The quality and uniqueness of walls built using only sub-soils cannot, arguably, be matched by cement-stabilised materials. The similarity with concrete is very often further exacerbated by the tendency to use finer soils, with lower clay and coarse gravel contents, producing a more uniform and fine-textured finish. Cement stabilisation also makes disposal and recycling of materials after use more problematic.

- **Stabilisation is not always necessary**
  In many situations, with careful selection of soil and attention to good design details, high-quality rammed earth buildings are achievable without the addition of cement. The use of cement and other stabilising additives may be seen as reducing the risk of material failure. However, it does not preclude the need for analysis of soil properties in preparation. On the whole the specifications for soil in stabilised rammed earth are more, not less, restrictive than those for general rammed earth.

D.3 Lime and other additives

Though the most widely used binder in stabilised rammed earth is cement, others include lime (hydraulic and non-hydraulic), bitumen and sodium silicate. Pozzolanic materials are normally used in combination with lime or cement to reduce additive content and improve material performance.

D.3.1 Lime stabilisation

Lime improvement and stabilisation of soils is a well established technique in civil engineering ground works such as road construction, where typically 1–3% (quick) lime is added to reduce soil moisture content, reduce plasticity and increase strength. In earthen construction lime (both non-hydraulic and hydraulic) is more typically used in proportions between 3 and 15% by mass.

Use of lime in rammed earth construction is much less common than use of cement. One of the few examples includes the uppermost, and noticeably lighter, layers in the AtEIC building walls at the CAT which have been lime stabilised (Figure 12). The benefits of adding lime are similar to those of adding cement, improved strength, durability and reduced shrinkage, though the rate of strength gain through carbonation of lime is much slower and lower than that using an equivalent amount of cement. Lime stabilisation is more tolerant of higher clay and organic matter contents (up to 20%), making it better suited as a stabiliser for more marginal soils less suited to rammed earth or cement stabilisation without granular stabilisation. The disadvantages are also similar to those of cement stabilisation. Though lower than for cement, carbon dioxide emissions in lime production are significant, so hydraulic lime should not be simply regarded as a more environmentally benign replacement for cement. Lime stabilised walls may require formwork support for longer periods (2 to 3 days), however.

Stabilisation with non-hydraulic and hydraulic lime, often in combination with cement, is ideally suited to soils with clay content in excess of that desirable for cement stabilisation alone. Generally, soils for lime stabilisation should meet the following criteria:

Sand + fine gravel content	5 to 70%
Silt content	10 to 60%
Clay content	20 to 60%
Plasticity index	20 to 30 (liquid limit 35 to 50)

Guidelines for lime stabilisation have been proposed based on soil plasticity[4]. Unsuitable soils include those with a combined clay and silt content of less than 30%, and those containing excessive sulfate and organic matter.

D.3.2 Other additives

Pozzolanic materials, used in conjunction with lime or cement, include ggbs, fly ash, crushed fire clay, rice husk ash and natural volcanic ash. Pozzolans can increase strength and durability, decrease reliance on cement or lime, and speed up strength gain.

Other additives have been used with varying degrees of success. Chemical resins, often developed for applications in road, concrete and other construction, have been used in Australia as both damp-proofing and protective coatings. Proprietary materials include acrylic resins, polymers, silane, siloxane and epoxy resins. Chemical agents are used both as additives during material preparation and, more widely, for protective coating of walls after construction.

Contact addresses

Rammed earth

Rowland Keable
In Situ Rammed Earth Co Ltd
86 Brougham Road
London, E8 4PB
(Rammed earth builder)

Jörge Depta
Lehmbauwerk
Willibald-Alexis-Strasse 15
10965 Berlin-Kreuzberg, Germany
(Rammed earth builder)

Martin Rauch
Lehm Ton Erde
Baukunst GmbH
Quadernstrasse 7
A-6824 Schlins, Austria
(Rammed earth builder/sculptor)

Dr Peter Walker
Dept Architecture & Civil Engineering
University of Bath
Bath, BA2 7AY
(Consultant/Researcher)

Andy Simmonds
Simmonds.Mills Architect-Builders
The Granary
Swinmoor Farm
Canon Bridge
Herefordshire, HR2 9JD
(Rammed earth consultant)

Stabilised rammed earth

Dr Matthew Hall
Centre for Built Environment
School of Environment and
 Development
Unit 9, Science Park
Sheffield Hallam University
City Campus
Howard Street
Sheffield, S1 1WB
(Researcher)

Bill Swaney
Hall Lane
Ashley
Market Harborough
Leicestershire, LE16 8HE
(Earth Structures (Europe) Limited)

(continued)

Earth building expertise

Cob Construction Company
Station Yard
Exminster
Exeter
Devon, EX6 8DZ
(Earth builders with experience of rammed earth)

Eartha
c/o Dirk Bouwens
Ivy Green
London Road
Wymondham, NR18 9JD

Tom Morton
Arc Architects
69 Burnside
Auchtermuchty
Fife, KY14 7AJ
(Architect with expertise in earth building)

Peter Trotman
Centre for Whole Life Construction
   and Conservation
BRE
Garston, Watford, WD25 9XX

Devon Earth Building Association
c/o P Child
South Coombe
Cheriton Fitzpaine
Crediton
Devon, EX17 4HP

Tim Hewitt
2 Thatched Cottage
Church Lane
Edingthorpe
N Walsham
Norfolk
(Earth builder with experience of stabilised rammed earth)

University of Plymouth
School of Architecture
Hoe Centre
Notte Street
Plymouth, PL1 2AR

Glossary of rammed earth terms

Additives: materials used to improve the physical properties of rammed earth.

Adobe: air-dried mud block, clay-lump or cob block.

Airborne sound insulation value (R_w): single quantity that characterises the airborne insulating qualities of a building element (BS 8233:1999[21]).

Bagging: rubbing onto the surface of a rammed earth wall to fill-in recesses and to prepare it for painting or rendering.

Binders: materials, including clay, cement and lime, used to fix soil particles together.

Blended or engineered soils: manufactured soils for rammed earth construction formed by combining various constituents to provide an ideal grading together with improved physical characteristics. Also sometimes referred to as 'granular stabilisation'. Blending offers the opportunity for greater use of otherwise unsuitable in-situ materials.

Block-out: special formwork, usually timber, used in rammed earth and stabilised rammed earth to form an opening or recess.

Boniness: sections of poorly compacted material caused by under-compaction or lack of fines.

Breathability: ability of a building material or element to allow water vapour to pass through it.

Cob: stacked mass mud–straw construction.

Cold joint: joint between successive layers of rammed earth or stabilised rammed earth caused by break in works.

Compaction: process of packing soil particles closer by removing air voids through manual or mechanical means.

Disaggregation: removal of larger aggregate particles during rammed earth construction to achieve smoother more uniform wall finish.

Drop test: simple method for estimating optimum moisture content of rammed earth and stabilised rammed earth.

Drying shrinkage: linear and volumetric reduction caused by loss of moisture from clay fraction on drying.

Durability: resistance to agents of decay, including water-borne deterioration and mechanical abrasion.

Dusting: loss of friable surface material.

Efflorescence: surface deposit of soluble salts, or calcium hydroxide in stabilised rammed earth, after drying.

Formwork: temporary support used during compaction. Also known as shuttering.

Granular stabilisation: see *Blended or engineered soils*.

'Hit and miss': process of rammed earth construction in which alternate wall panels are built initially and then in-filled on a 'second pass' following some drying of the initial panels.

Hygroscopic: having the characteristic of absorbing moisture from the atmosphere.

Maximum dry density: the maximum dry density for densest possible packing of particles of rammed earth (BS 1377-1:1990[26]).

Moisture content: water content represented as a percentage of the dry mass of solid materials.

Movement joint: vertical joints between panels of rammed earth to accommodate moisture and thermal movements.

Optimum moisture content: the moisture content of loose rammed earth or stabilised rammed earth material at which the specified compaction method will achieve the maximum dry density (BS 1377-1:1990[26]).

Plucking: surface damage to rammed earth that occurs during striking of the formwork, caused by adhesion with the formwork panel.

Pozzolans: materials used to develop hydraulic set and rate of hardening in lime. Used as a cement replacement material. Materials include fly ash, ground granulated blastfurnace slag, ground brick dust and volcanic ash.

Glossary

Protective coating: application of coating material to enhance abrasion or water resistance.

Rammed earth: monolithic earth-based material using sub-soils free from additives and compacted between temporary formwork.

Rammed chalk: monolithic wall material composed of crushed chalk and compacted using the rammed earth technique.

Rammer: manual or mechanical tool used in compaction of rammed earth.

Stabilisation: the process of binding together sub-soil particles through the addition of a hydraulic binder such as cement or hydraulic lime or a non-hydraulic binder such as hydrated lime.

Shuttering: see *Formwork*.

Stop-end: end piece of formwork placed across thickness of wall.

Surface mass (m'): mass per unit surface area of a single leaf wall (BS 8233:1999[21]).

Thermal mass: dense material, such as rammed earth, that stores and releases heat energy.

Tooling: surface decoration of rammed earth wall; often used to disguise small defects, imperfections and joints.

References

[1] **Easton D** (1996). *The rammed earth house.* Chelsea Green Publications Company, Vermont, USA.

[2] **Middleton G F** (1952). Earth-wall construction. Pisé or rammed earth; adobe or puddled earth; stabilised earth. *Bulletin* No 5, Department of Works and Housing, Sydney, Australia.

[3] **Standards Australia** (2002). *The Australian earth building handbook.* Sydney. 151 pp.

[4] **Houben H and Guillaud H** (1994). *Earth construction: a comprehensive guide.* Intermediate Technology Publications, London, UK.

[5] **Webb D** (1995). Performance consideration for the use of stabilised soil building blocks with cob construction in the United Kingdom. *Out of Earth II Conference Proceedings, University of Plymouth, 1995.* pp 90–117.

[6] **Pearson G T** (1992). *Conservation of clay and chalk buildings.* Donhead, Shaftesbury, UK.

[7] **Williams-Ellis C, Eastwick-Field J and Eastwick-Field E** (1916). *Building in cob, pisé and stabilised earth.* Cambridge University Press, 1916; republished by Donhead, Shaftesbury, 1999.

[8] **Kapfinger O** (2001). *Martin Rauch: Rammed Earth _Lehm und Architektur_ Terra cruda.* Birkhäuser, Basel, Switzerland.

[9] **Minke G** (2000). *Earth construction handbook. The building material earth in modern architecture.* WIT Press, Southampton, UK.

[10] **Rauch M** (2005). Contemporary earth projects without stabilisation. *Earthbuild 2005 conference abstract,* University of Technology, Sydney, Australia, January 2005.

[11] **Centre for Alternative Technology, AtEIC.** *Autonomous Environmental Information Centre. Factsheet,* 2000. Centre for Alternative Technology Publications, Machynlleth, UK.

[12] **ODPM.** *The Building Regulations 2000.* Statutory Instrument 2000 No 2531. The Stationery Office, London, 2000.

[13] **ODPM.** *The Building Regulations 2000. Approved Document C. Site preparation and resistance to contaminants and moisture.* 2004 edition. The Stationery Office, London, 2004.

[14] SAP 2001. *The Government's Standard Assessment Procedure for Energy Rating of Dwellings.* Revised edition with minor corrections 20 December 2002. Garston, BRE. See website http://projects.bre.co.uk/sap2001; the latest draft of SAP2005 can be seen at: www.bre.co.uk/sap2005

[15] **Palmgren L A** (2003). *Svenska jordhus av lera eller kalk.* KTH Stockholm.

[16] **British Standards Institution.** Methods of test for soils for civil engineering purposes. Part 2. Classification tests. *British Standard* BS 1377-2:1990. BSI, London, 1990.

[17] **British Standards Institution.** Methods of test for soils for civil engineering purposes. Part 4. Compaction-related tests. *British Standard* BS 1377-4:1990. BSI, London, 1990.

[18] **British Standards Institution.** Methods of test for soils for civil engineering purposes. *British Standard* BS 1377:1990. BSI, London, 1990.

[19] **Building Research Establishment** (1986). Rising damp in walls: diagnosis and treatment. *BRE Digest* 245. BRE Bookshop, Garston.

[20] **Middleton G F** (1987). (Revised by Schneider L M, 1992). *Bulletin 5. Earth Wall Construction.* Fourth edition. CSIRO Division of Building, Construction and Engineering, North Ryde, Australia.

[21] **British Standards Institution.** Code of practice for sound insulation and noise reduction for buildings. *British Standard* BS 8233:1999. BSI, London, 1999.

[22] **American Society for Testing and Materials.** Standard test method for measurement of masonry flexural bond strength. ASTM C1072.

[23] **Centre for the Development of Enterprise** (2000). *Compressed earth blocks: testing procedures.* CDE, Brussels, Belgium.

[24] **Standard New Zealand.** Engineering design of earth buildings. *New Zealand Standard* NZS 4297:1998. Standard New Zealand, Wellington, New Zealand.

[25] **Pritchett I** (2004). The modern face of lime. *Building for a Future,* **14** (2) 62–63, Autumn 2004.

[26] **British Standards Institution.** Methods of test for soils for civil engineering purposes. Part 1. General requirements and sample preparation. *British Standard* BS 1377-1:1990. BSI, London, 1990.

Bibliography

A rammed earth house in Massachusetts. *APT Bulletin,* 1983, **15** (2) 33–37.

Alley P J (1948). Rammed earth construction. *New Zealand Engineering,* June 10, 582.

American Society for Testing and Materials (1989). Standard test methods for freezing and thawing compacted soil-cement mixtures. ASTM D 560-89. ASTM Standards, Philadelphia, USA.

Armstrong S (1988). Pied à terre. *New Scientist,* 10 March 1988, pp 60–64.

Borer P and Lea D (2000). Rammed earth first for alternative technology HQ. *The Architects' Journal,* **212** (8) 16.

Building Design (1999). Raw-earth centre on site; Architects: Pat Borer. 6, 1418, Nov 12th.

Building Research Establishment (1996). Earth building: a topic of continuing interest. In: *75 years of building research.* BRE Bookshop, Garston, Watford.

Clarke R (2001). How energy efficient are rammed earth walls? *Chartered Building Professional,* February 2001, 14.

Crowley M (1997). Quality control for earth structures. *Australian Institute of Building Papers* 8, 109–118.

CSIRO Division of Building, Construction and Engineering (1996). *Earth wall construction.* Building Technology File Number Twelve. CSIRO Australia, North Ryde, Australia.

CSIRO Division of Building, Construction and Engineering (1996). *Pisé (rammed earth) construction.* Building Technology File Number Thirteen. CSIRO Division of Building, Construction and Engineering, CSIRO Australia, North Ryde, Australia.

Dayton L (1991). Saving mud monuments. *New Scientist,* 25 May 1991, pp 38–42.

Dethier J (1982). *Down to earth.* Thames & Hudson, London.

Dethier J (1985). Story of pisé. *Architectural Review,* 178, 1064, Oct, 67–68.

Department for Transport, Local Government and the Regions/Department for Environment, Food and Rural Affairs. *Limiting thermal bridging and air leakage: robust construction details for dwellings and similar buildings.* London, The Stationery Office, 2001.

Devon Earth Building Association (1996). *Cob and the building regulations.*

Earth Building Association of Australia (2001). *Earth building book. Draft for Comment.* Draft Code 05/01. Earth Building Association of Australia, Wangaratta, Australia.

Elizabeth L and Adams C (2000). *Alternative construction. Contemporary natural building methods.* John Wiley & Sons, New York, USA.

Fromonot F (1997). Ateliers a Tuscon [Artists' studios and residences in Tuscon]; Architects: Rick Joy. *Architecture D'Aujourd'Hui,* 312, September, 48–51.

Gommersall F (2002). Earth construction. *Building for a Future,* **12** (3).

Hall M (2002). Rammed earth: traditional methods, modern techniques, sustainable future. *Building Engineer,* Nov, 22–24.

Hall M and Djerbib Y (2004). Moisture ingress in rammed earth: Part 1 – The effect of soil particle size distribution on the rate of capillary suction. *Construction and Building Materials,* **18** (4) 269–281.

Hall M and Djerbib Y (2004). Rammed earth sample production: context, recommendations and consistency. *Construction and Building Materials,* **18** (4) 282–286.

Hannay P (2000). Ground force: Architects: Pat Borer and David Lea. *RIBA Journal,* **107** (11) 34–40.

Harries W J R, Clark D and Watson L (2000). A rational return to earth as contemporary building material. *Terra 2000, 8th International Conference on the Study and Conservation of Earthen Structures,* Torquay, Devon.

Hladik M (2002). Rammed earth: Martin Rauch. *Architecture Aujourdhui.* 339, 10–11.

Hughes R (1983). Material and structural behaviour of soil constructed walls. *Momentum,* pp 175–188.

Hurd J and Gourley B (2000). *Terra Britannica. A celebration of earthen structures in Great Britain and Ireland.* James & James (Science Publishers) Ltd, London, UK.

Keable J (1994). *Rammed earth standards.* ODA Final Project Report R 4864C.

Keable J (1996). *Rammed earth structures. A code of practice.* Intermediate Technology Publications, London, UK.

Keable R (2001). East of Eden. *Building for a Future,* **10** (4).

Kennedy J F, Bates A, Wanek C and Smith M (2002). *The art of natural building.* New Society Publications.

King B P E (1996). *Buildings of earth and straw. Structural design for rammed earth and straw bale architecture.* Ecological Design Press, Sausalito, California, USA.

Lehmbau Regeln. Begriffe; Baustoffe; Bauteile. (1999) Friedr. Vieweg & Sohn Verlagsgesellscahaft mbH, Braunschweig/Wiesbaden, Germany.

Bibliography

Lilley D M and Robinson J (1995). Ultimate strength of rammed earth walls with openings. *Proceedings of the Institution of Civil Engineers, Structures and Buildings,* August, 110, 278–287.

Little B and Morton T (2001). *Building with earth in Scotland. Innovative design and sustainability.* Scottish Executive Central Research Unit, Edinburgh, UK.

Maiti S K and Mandal J N (1985). Rammed earth house construction. *Journal Geotechnical Engineering,* **111** (11) 1323–1328.

McCann J (1995) *Clay and cob buildings.* Shire Publications.

McHenry P G (1984). *Adobe and rammed earth buildings. Design and construction.* A Wiley-Interscience Publication, New York, USA.

Meade M and Garcias J C (1985). Return to earth. *Architectural Review,* No 63/10.

Merrill A F (1979). *The rammed earth house.* Ann Arbor, London.

Middleton G F (1953). *Build your house of earth. A manual of pisé and adobe construction.* Angus and Robertson, Sydney, Australia.

Miller L and Miller D (1980). *Manual for building a rammed earth wall.* Rammed Earth Homes, c/o L Miller and D Greeley, USA.

Miller L and Miller D (1982). *Rammed earth. A selected bibliography with a world overview.* Rammed Earth Institute International, Colorado, USA.

Ministerio de Obras Públicas y Transportes (1992). *Bases Para el Diseña y Construccion Con Tapial,* Centro de Publicaciones, Secretaria General Tecnica, Ministerio de Obras Públicas y Transportes, Madrid, Spain.

Moderner Lehmbau (2002). *Internationale Beiträge zum modernen Lehmbau,* KirchBauhof GmbH und bei den Autoren, April 2002, Berlin Germany.

Montgomery D (1998). *How does cement stabilisation work?* Stabilised Soil Research Progress Report SSRPR02. Coventry: Development Technology Unit, School of Engineering, University of Warwick, October.

Morris H W (1993). The strength of engineered earth buildings. *IPENZ Annual Conference, Sustainable Development, Hamilton, 5–9 February.* 660–671.

Mukerji K (1989). *Rammed earth construction.* United Nations ESCAP, Building Technology Series, No 1.

Norton J (1997). *Building with earth. A handbook.* Second edition, Intermediate Technology Publications, London, UK.

Out of Earth Conference Proceedings 1994. Eds Watson and Harding. University of Plymouth.

Out of Earth II Conference Proceedings 1995. Eds Watson and Harries. University of Plymouth.

Simmonds A and Keable R (1992). Down to earth. *Building for a Future,* **2** (2).

Sinha S and Schmann P (1994). Earth building today. *The Architects' Journal,* 22 September.

Smith E W and Austin G S (1996). *Adobe, pressed-earth, and rammed-earth industries in New Mexico* (revised edition). Bulletin 159. New Mexico Bureau of Mines and Mineral Resources, Socorro, USA.

Spence R J S and Cook D J (1982). *Building materials in developing countries.* Wiley.

Standard New Zealand. Materials and workmanship for earth buildings. *New Zealand Standard* NZS 4298:1998. Standard New Zealand, Wellington, New Zealand.

Standard New Zealand. Earth buildings not requiring specific design. *New Zealand Standard* NZS 4299:1998. Standard New Zealand, Wellington, New Zealand.

Standards Association of Zimbabwe (2001). *Rammed earth structures.* SAZS 2001 724.

Steingass P (2000). *LEHM 2000,* Overall Verlag Berlin, Berlin, Germany.

Steingass P and Vietzen W (1999). *Moderner Lehmbau Einführung und Messekatalog 1999,* Overall Verlag Berlin, Berlin, Germany.

Stulz R and Mukerji K (1988). *Appropriate building materials.* SKAT Publications, Switzerland.

Terra 2000, 8th International Conference on the Study and Conservation of Earthen Architecture. Postprints. English Heritage, May 2000, Torquay, Devon.

Terra 2000, 8th International Conference on the Study and Conservation of Earthen Architecture. Preprints. English Heritage, May 2000, Torquay, Devon.

Thompson F (1996). Earthly fortresses + Unknown communal rammed-earth fortresses in southern China. *Architecture Review,* **199** (1188) 84–86.

Tibbets J M (2001). Emphasis on rammed earth – the rational. *Interaméricas Adobe Builder,* **9** 4–33.

Venkatarama B V and Jagadish K S (1993). The static compaction of soils. *Géotechnique,* **43** (2) 337–341.

Voelcker A (2002). Rammed earth: Martin Rauch. *Architecture Review,* **211** (1263) 97.

Walker P (2000). Review and experimental comparison of erosion tests for earth blocks. *Proceedings 8th International Conference on the Study and Conservation of Earthen Architecture, Torquay, May 2000.* pp 176–181.

Walker P, Ayala R and Dobson S (2002). Reinforced composite rammed earth in flexure. In: *Proceedings 3rd International Conference on Non-Conventional Materials & Technologies, Hanoi.* pp 439–445.

Walker P and Dobson S (2001). Pullout tests on deformed and plain rebars in cement-stabilised rammed earth. *Journal of Materials in Civil Engineering,* **13** (4) 291–297.

Warren J (1999). *Conservation of earth structures.* Butterworth-Heinemann, Oxford.

Weiner P (1996). Rammed-earth compound. *Fine Homebuilding,* 101, 90–95.

Zukunft Lehmbau (2002). Conference, Symposium and Annual General Meeting. Dachverband Lehm e.V./Bauhaus-Universität Weimar.

Zur Nieden G and Ziegert C (2002). *Neue Lehn-Hauser international: projektbeispiele, konstruktion, details.* Bauwerk Verlag.

Index

Page numbers in *italics* refer to illustrations.

abrasion 86, *87*
   internal walls 18
   protection against 42, 88, 115
   sampling 101
   specifications 100, 111, 115
   stabilised rammed earth 127
   testing resistance to 42, 109–10, *110*
   see also durability; dusting; erosion
acoustic properties and performance 75, 81
   Building Regulations 26
   movement joints 26, 56, 75
   shrinkage and 40–1
   walls 17, 18, 81
additives 34–5
   in furniture, sculptures, etc 21
   maintenance and repair work 94
   soil constituents and 30, 33
   specification 113
   stabilised rammed earth 129–30
   test specimens 103
addresses 131–2
aggregates
   colour effects 31
   costs 35
   granular stabilisation 35–6
   maximum size 103, 113
   soil density and moisture content 38
   stabilised rammed earth 126
air
   in soils 30
   see also density
Alabama, house *9*
Alhambra, Granada *4*
altars *8*, 21
architectural aspects 11
   see also design; details
arsenic 30
Ashley, stables *125*
AtEIC Building *6*, 36, 129
Australia, house *9*

bending and compression, design for 120–1
Bird-in-Bush Nursery *7*
blended soils *see* granular stabilisation
block-outs 24, 51
bolts *see* fixings
boniness *91*
   abrasion testing 109
   compressive strength and 39
   fixings and 75
   particle size and 37
   repair 94
   specifications 89, 90, 100, 116

Brandenburg *8*
Brimington Bowls Club Pavilion *125*
Building Regulations 21, 24–7, 64, 81

cantilevered formwork 46
cement *see* stabilised rammed earth
Centre for Alternative Technology *6*, 36, *97*, 129
ceramic tiles 73
chalk structures 5, *5*, 20, 21, 30–1
Chapel of Reconciliation, Berlin *3*, *8*, 34, 36
Chelsea Flower Show *6*
cladding 14, 17, 19, 69, 126
clamped formwork 46
coatings *see* protective coatings
colour
   control of 11, 31, 59
   matching for repair work 90, 93–4
   natural soil and aggregates 31
   protective coatings and 31, 70
   specification 89, 90, 100, 116
   variation *91*
compaction *32*, 33–4, *33*, 53–5, *54*
   flexural and shear strength 39
   natural fibre additives 35
   quality control 15
   specification 115
   testing 102
   see also heavy manual compaction test
compliance tests 99–101
compression and bending, design for 120–2
compressive strength 80, 83, 120
   particle size and 37
   rammed chalk 31
   specification 100, 113
   stabilised rammed earth 125, 127
   testing 39, 101, 103–4
concentrated compression loads 121–2
conduits 74
   see also electrical services
construction *2*, 13, 15–16, 22–4, *23*, 42, 45–59
   see also Building Regulations; specification; tolerances
contact addresses 131–2
contamination 30, 52
contracts 27, 59
   see also specifications
corners 13, 23, 56, 86
costs 15–16, 19, 35, 128
covers 23–4, 46, 58, 86, 116
   see also protection
cracking 39–41, 87–8, 89, 90, *92*, 94
   see also deformation; plasticity; shrinkage
creep 18, 39

damage 85–8
    see also abrasion; damp-proofing; durability; erosion; maintenance; protection; rainfall damage; splash damage
damp-proofing 42, 62–4, *62*
    Building Regulations 26
    floors 21
    maintenance 87, 89
    specification 111, 112
    see also moisture resistance
deformation 39–40, 84
    see also cracking; movement joints; plasticity
density 32, 79
    additives and 33
    chalk 31
    moisture content and 34
    specification 100, 115
    testing 38, 101–2
design 17–27, 79–84, 119–23
    durability 13
    formwork and 15, 22
    surface finishes 11
desk *21*
details 61–78
Dragons Retreat *10*
drainage 42, 64
drop tests 51, 102
dry density see density
drying
    colour changes 59
    influences on 40, 58, 72
    plasticity and 33
    propping during 22, 24, 40, *41*
    protection during 24, 115, 116
    rate of 40, 72, 100
    see also shrinkage
ductwork 24
durability 13–14, 41–2, 61, 126, 127
    see also abrasion; erosion; protection; weathering
dusting 93
    minimising 13, 18, 42, 70, 88
    specification 90
    stabilised rammed earth 127
    see also abrasion; friability

eaves 13, 42, 69
Eden Project *2, 6*
efflorescence 21, 30, 90, *93*, 94, 100
    see also soluble salts
elastic deformation 84
elastic modulus 40, 80, 84
electrical services 24, 74, *74*
energy efficiency 10, 14, 26–7, 43
    see also environmental aspects; thermal properties and performance
engineering design 79–84
environmental aspects 10–11, 35, 125–6, 128, 129
    see also energy efficiency; thermal properties and performance
erosion 86, *86*
    maintenance and repair 89, 94
    protection 71
    rate of 42
    sampling 101
    specification 100, 111
    testing resistance to 42, 106–9, *108*
    see also abrasion; durability; splash damage; weathering
external walls 17–18
    durability 13
    protection 42, 69, 70, 71, 88
    protective coatings 26, 61, 69
    thermal properties and insulation 14, 17–18, 27, 71, 76–7
    see also damp-proofing

fire precautions 25–6, 41, 43, 56, 81
fireplaces 21
fixings 40, 57, 68, 75, 111, 118
    see also through-bolted formwork
flashings 89, 112
flexural loads, design for 123
flexural tensile strength 39, 80, 105–6
flood protection 13, 25, 41, 42, 61, 87
floors 20–1, 26, 33
    see also underfloor heating
footings 61–4, 94, 127
formwork 46–50
    cost control 15–16
    defects caused by 89, 90, 94
    during construction 16, 22, 45, 46, 55–6
    openings 51, 65
    removal 55, 59, 116
    services 74
    specification 115, 116–17
    surface finishes and 59, 90, *92*, 116
    tie holes 117
    tolerances 78, 117
    see also block-outs
foundations see footings
free-standing walls 18, 82
friability 32, 37, 90
    see also dusting
furniture 21

Genesis Project 95
glossary 133–5
grading, soils 29, 31–2
    classification tests 36
    material selection 14, 37–8
    specification 100, 113
    stabilised rammed earth 126, 130
granular stabilisation 15, 35–6

hardness testing 100
health and safety 24
heavy manual compaction test 33, 38, 99, 102, 113
'hit and miss' construction 56
humidity regulation 12–13, 18, 43, 70

inspection 112
internal walls 13, 18, 70, 71, 81

Jasmine Cottage *10*
joints 56, 112, 115, 116
    see also movement joints

Kindersley Centre *7*

lime additives 129–30
lime renders 17, 70, 71, *73*

Index

lintels 51, 65, 66, 68

maintenance 13, 21, 70, 85–94, *92*, 118
materials 29–43
   Building Regulations 27
   protective coatings 70
   repairs 93–4
   sources 11, 15, 30, 35, 36, 93, 114
   see also additives; chalk; soils; testing
maturation 51
mixing and mixers 35, 51–3, *52*, *53*, 113
moisture content 30
   additives and 33
   aggregates and 38
   compaction and 33, 55
   compressive strength and 39
   design and 25, 79
   dry density and 34
   elastic deformation 84
   increases in 41, 51
   specification 100, 113
   testing 38, 51, 101–2, 113
   see also drying; plasticity; shrinkage
moisture ingress 41–2
   see also flood protection; rainfall damage; rising damp; splash damage
moisture resistance 26, 34, 127
   see also damp-proofing
Morocco 4
Mount Pleasant Ecological Business Park *7*, *20*
movement joints 40, 41, 56–7, *57*, *73*, 78
   acoustic protection 26, 56, 75
   airtightness 75
   fire protection 25, 56
   maintenance 89
   shrinkage 40, 56, 73
   spacing 41, 84
   specification 117
moving formwork 46–7, 55, 56

natural fibre additives 34–5, 43

office desk *21*
openings 14, 51, 65–8, *65*, 83
organic matter 30, 36, 37, 113, 126
out-of-plane flexural loads 123

partial safety factor 39, 83, 119
particle sizes 37, 101
   see also grading
patching *see* maintenance
placing 24, 52–3, 115
plasticity 31, 32–3, 36, 37
   see also cracking; moisture content
plucking 59, 90, *92*
pozzolans 128, 129, 130
prefabricated rammed earth 15, 19, *19*, 94, 99
productivity 22, 47, 50, 55
   see also speed of construction
propping 22, 24, 40, *41*, 57
protection 13–14, 23–4, 42, 58–9, *58*, 87
   against abrasion 42, 88, 115
   against erosion 71
   external walls 42, 69, 70, 71, 88
   floors 21

free-standing walls 18
internal walls 13, 18, 70, 71
specification 27, 115–16
see also eaves; fire precautions; flood protection; protective coatings; roofs
protective coatings 70–3
   colour and 31, 70
   external walls 26, 61, 69
   floors 21
   internal surface 13, *72*
   maintenance checks 89
   problems 13–14, 21, 26, 43, 70, 71, *71*
   shrinkage and 24, 70, 72–3
   see also covers; renders and rendering

quality control 15, 27
   see also testing

radon barriers 62
rainfall damage 42, 85–6, *85*, 88, 127
   see also covers; moisture ingress; splash damage
rammed earth
   advantages 10–13
   description and uses 2–3, 96
   history and development 3–10, 95–7
   limitations 13–16
   properties 79–81, 99–110
   see also stabilised rammed earth
rammers and ramming *see* compaction
renders and rendering 17–18, 24, 70, 71–2, 94
   see also protective coatings
repairs *see* maintenance
retaining walls 18–19
Rhone Valley 4
rising damp 26, 42, 87
   see also damp-proofing; moisture ingress
roofs 23, 58–9, 69

safety 24
sampling 36, 100, 101, 114
sculpture 21
services 24, 74
   see also drainage; electrical services; underfloor heating; ventilation
shear, design for 123
shear strength 39, 80, 106, *107*
sheeting *see* covers
shrinkage
   acoustic properties and 40–1
   construction sequencing and 55–6
   design for 24, 80, 84
   internal walls 18
   movement joints 40, 56, 73
   natural fibre additives and 34
   openings 65–6, 68
   plasticity and 33
   protective coatings and 24, 70, 72–3
   sampling 101
   soil selection 37
   specification 100, 114
   testing 40–1, 100, 104–5
shuttering *see* formwork
site investigation 36
sitework *see* construction
skid steer loaders *52*, *53*

slenderness 14, 82, 120, 121
soil characteristics 31–3
    see also colour; density; grading; moisture content; plasticity; soil classification tests
soil classification tests 36–7
soil surveys 36
soils 29
    additives' effects 30, 33
    performance specification 100
    sampling 101
    selection 35–8
    stabilised rammed earth 126
    storage and preparation 15, 22, 45–6, 51, 113
    suitability 14–15
    see also grading; materials; mixing; testing
soluble salts 30
    specification 100, 113
    stabilised rammed earth 126
    testing for 36, 37
    see also efflorescence
Somerset College of Arts and Technology 95
sound see acoustic properties and performance
specifications 89–90, 99–100, 111–18
speed of construction 13
    see also productivity
splash damage 64, *64*, 86, 115
stabilised rammed earth 1, *9*, 125–30, *125*
    compressive strength 39, 103
    floors 21
    history and development 9–10, 95
    mixing 52
    retaining walls 18–19
    testing 38, 103
stables, Ashley *125*
static formwork 47, *48*, 55, 56
stoves 21
straw additives 34–5
strength 13, 34, 127
    see also compressive strength; flexural tensile strength; shear strength
striking see formwork, removal
structural design 81–3, 119–23
structure, Building Regulations 25
stud walling 18
support see propping
surface finishes *12*, 59
    formwork 59, 90, 116
    sampling 101
    services installations 74
    specification 89, 100, 116–17
    tooling 11, *12*, 90, 94
    see also colour; placing; protective coatings
surface hardness testing 100
sustainable construction see environmental aspects

temperature, during compaction 115
test panels 27, 45, 70, 109, 110
testing 25, 35, 38–43, 99–110
    arsenic concentrations 30
    moisture content 38, 51, 101–2, 113
    soluble salts 36, 37
    specification 112, 114
    textural variation 70
    see also drop tests; heavy manual compaction test; quality control; soil classification tests

textural variation 11, *91*
    grading and 32
    maintenance and repair 94
    specification 89, 90, 100, 116
    testing 70
    see also formwork, surface finishes
thermal expansion 41
thermal insulation 75
    external walls 14, 17–18, 27, 71, 76–7
    stabilised rammed earth 126, 127
thermal properties and performance 13, 80
    Building Regulations 26–7
    natural fibre additives 34
    testing 43
    walls 14, 17–18, 27
    see also energy efficiency
thickness
    render coats 71
    test specimens 107, 109
    tolerances 78, 117
    walls 2, 14, 81, 82
through-bolted formwork 46, 50, *50*
tie holes 117
ties see fixings
tiles 73
timber formwork 47, 49–50, *49*, 55
timber frame construction 14, 18
tolerances 78, 117
transportation 11, 15, 35, 115

U-values see thermal properties and performance
underfloor heating 21, 74

ventilation 70, 75

wall bases see footings
wall plates 68, 111, 118, 127
water damage 86, *86*, 87
    see also moisture ingress; rainfall damage; splash damage
water pipes 74
weathering 85–8
    see also erosion; protection; rainfall damage; splash damage
Weilburg 4
Winchester, chalk houses 5
WISE Project *96*
Woodley Park Sports Centre *6*, *72*

Zeesen 8